Quantum Key Distribution Networks

Miralem Mehic • Stefan Rass •
Peppino Fazio • Miroslav Voznak

Quantum Key Distribution Networks

A Quality of Service Perspective

 Springer

Miralem Mehic
Department of Telecommunications,
Faculty of Electrical Engineering
University of Sarajevo
Sarajevo, Bosnia and Herzegovina

Stefan Rass
Secure Systems Group, LIT Secure and
Correct Systems Lab
Johannes Kepler University
Linz, Austria

Peppino Fazio
Department of Telecommunications
VSB-Technical University of Ostrava
Ostrava, Czech Republic

Miroslav Voznak
Department of Telecommunications
VSB-Technical University of Ostrava
Ostrava, Czech Republic

ISBN 978-3-031-06610-8   ISBN 978-3-031-06608-5   (eBook)
https://doi.org/10.1007/978-3-031-06608-5

This Springer imprint is published by the registered company Springer Nature Switzerland AG
The registered company address is: Gewerbestrasse 11, 6330 Cham, Switzerland

Miralem: to Lejla and my family
Stefan: dedicated to my loving family
Peppino: to my lovely family, mom, dad, and
Francesco
Miroslav: to my beloved family

Acknowledgments

The research leading to the published results was supported by the Ministry of the Interior of the Czech Republic under grant ID VJ01010008 within the project Network Cybersecurity in Post-Quantum Era.

We would like to thank Oliver Maurhart, Marcin Niemiec, and Emir Dervisevic for helpful discussions and comments on the manuscript.

Contents

Acronyms

5G	The fifth generation of cellular networks
AAU	Active Antenna Unit
AES	Advanced Encryption Standard
AIT	Austrian Institute of Technology
API	Application Programmers Interface
ASMT	Arbitrarily Secure Message Transmission
ATM	Asynchronous Transfer Mode
BBN	Bolt Beranek and Newman
BBU	Base Band Unit
BF	Bellman-Ford
BGP	Border Gateway Protocol
CAC	Call Admission Control
CC	Common Criteria
CIA	Confidentiality-Integrity-Availability
CLI	Command Line Interface
CO	Central Office
CV-QKD	Continuous-Variable QKD
CVSS	Common Vulnerability Scoring System
DDoS	Distributed Denial-of-Service
DH	Diffie-Hellman key agreement primitive
DHE	Ephemeral Diffie-Hellman (DHE)
DiffServ	Differentiated Services
DIQKD	Device-Independent Quantum Key Distribution
DoS	Denial-of-Service
DSCP	Differentiated Services Code Point
DSDV	Destination-Sequenced Distance-Vector
DU	Digital Unit
DV	Distance Vector
DV-QKD	Discrete Variables QKD
E2E	End-to-End
ECN	Explicit Congestion Notification

eCPRI	Enhanced Common Public Radio Interface
ERO	Explicit Route Object
ESP	Encapsulating Security Payload
FEC	Forwarding Equivalence Class
Fi-Wi	Fiber/Wireless
FQKD	Flexible QoS model for QKD Networks
FR	Frame Relay
GMPLS	Generalized Multi-Protocol Label Switching
GPL	GNU Public License
GPSR	Greedy Perimeter Stateless Routing in Wireless Networks
GPSRQ	Greedy Perimeter Stateless Routing Protocol for QKD Networks
GUI	Graphical User Interface
HARQ	Hybrid Automatic Retransmit reQuest
HMAC	Hash Message Authentication Code
HOM	Hong-Ou-Mandel
IETF	Internet Engineering Task Force
IKE	Internet Key Exchange
IntServ	Integrated Services
IoT	Internet of Things
IPComp	IP payload compression protocol
IPsec	Internet Protocol security
ISAKMP	Internet Security Association and Key Management Protocol
IS-IS	Intermediate System to Intermediate System
ISP	Internet Service Provider
ITS	Information-Thoeretic Security
IV	Initialization Vector
KMS	Key Manager System
KSID	Key_Stream_ID
LDPC	Low Density Parity Check
LER	Label Edge Routers
LKMS	Local Key Manager System
LP	Linear Program
LS	Link-State
LSA	Link-State Advertisement
LSP	Label Switch Path
LSU	Link-State Update
MAC	Message Authentication Code
MACsec	Media Access Control security
MANET	Mobile Ad Hoc Network
MANO	Management and Orchestration
MDI-QKD	Measurement Device Independent Quantum Key Distribution
MGSS	Multi-Goal Security Strategy
MPLS	Multi-Protocol Label Switching
MPT	Multipath Transmission
MSS	Maximum Segment Size

MTU	Maximum Transmission Unit
NAT	Network Address Translation
NFV	Network Function Virtualization
NIST	National Institute of Standards and Technology
NR	New Radio
NVD	National Vulnerability Database
OLA	Operational Level Agreement
OS	Operating System
OSPF	Open Shortest Path First
OTP	One-Time Pad
P2MP	Point-to-MultiPoint
P2P	Point-to-Point
PCE	Path Computation Element
PCEP	Path Computation Element Protocol
PER	Provider Edge Router
PFS	Perfect Forward Secrecy
PITM	Person-in-the-Middle
PKI	Public Key Infrastructure
PON	Passive Optical Network
PPP	Point-to-Point Protocol
PSMT	Perfectly Secure Message Transmission
Q3P	Quantum Point-to-Point Protocol
QBER	Quantum Bit Error Rate
QKD	Quantum Key Distribution
QKDNetSim	QKD Network Simulator
QKRA	Quantum Key Reservation Approach
QoS	Quality of Service
QPFS	Quantum Perfect Forward Secrecy
QRNG	Quantum Random Number Generator
QSIP	QKD Signaling Protocol
QUANTUM5	Quantum Cybersecurity in 5G Networks
RAN	Radio Access Network
RAT	Radio Access Technology
REST	REpresentational State Transfer
RIP	Routing Information Protocol
RRH	Remote Radio Head
RRU	Remote Radio Unit
RSVP	Resource Reservation Protocol
RTT	Round-Trip Time
RU	Radio Unit
SAD	Security Association Database
SAE	Secure Application Entity
SDN	Software Defined Networking
SECOQC	Secure Communication based on Quantum Cryptography
SeQKEIP	Secure Quantum Key Exchange Internet Protocol

SIP	Session Initiation Protocol
SKEYID	Session Key ID
SLA	Service Level Agreement
SPAD	Single-Photon Avalanche Diodes
SPD	Security Policy Database
SPD	Single-Photon Detector
SPI	Security Parameter Index
TCP	Transmission Control Protocol
TLS	Transport Layer Security
TTL	Time to Live
TVA	Topological Vulnerability Analysis
UDP	User Datagram Protocol
UE	User Equipment
URI	Uniform Resource Identifier
VANET	Vehicular Ad Hoc Network
vCPE	Virtual Customer Premise Equipment
vEPC	Virtual Evolved Packet Core
VoIP	Voice over IP
VPN	Virtual Private Network
vRAN	Virtual Radio Access Network
VSB	Technical University of Ostrava
WDM	Wavelength Division Multiplexing

List of Symbols

$\{0, 1\}^n$	Set of bitstrings of length n
$\{0, 1\}^*$	Set of bitstrings of any length $(= \bigcup_{n \in \mathbb{N}} \{0, 1\}^n)$
$\|m\|$	Length of bitstring $m \in \{0, 1\}^*$
$\oplus$	Bitwise XOR between two strings
π	Path, represented as ordered subset $\pi \subseteq E$ of edges in a graph $G = (V, E)$
$V(\pi)$	Nodes along a path π
$\|\pi\|$	Length of a path π as number of edges (hops)
$\mathscr{A}$	Adversary structure; a family (set) of sets
$\mathscr{P}(X)$	Power-set of the set X
$\mathbb{Z}_p$	Finite field of prime order p, with modulo arithmetic

Chapter 1
Fundamentals of Quantum Key Distribution

Establishing secure cryptographic keys through an untrusted network is a fundamental cryptographic task [1]. While the use of public key infrastructure based on computational intractability assumptions prevail, these solutions remain theoretically breakable. They are under constant threat as computational power continues to increase, new algorithms are discovered and new computing architectures become available [2]. To establish information-secure cryptographic keys which carry no such risks, quantum information theory proposes reliance on the laws of physics with the use of Quantum Key Distribution (QKD) [3].

Quantum cryptography makes use of photons (particles of light), benefiting from some of their properties to act as information carriers. Specifically, where normal communication uses electrical voltage levels or laser pulses to encode 0 or 1 bits, quantum communication uses light particles as carriers. It has become common practice to use Dirac's notation $|\psi\rangle$ (see [4]) to denote a light particle ψ. This particle is called a *qubit* and has information encoded in it by placing ψ into a selected polarization plane, either horizontal $\leftrightarrow$, vertical $\updownarrow$ or a combination of the two. Consequently, the general particle is represented as a linear combination of these two basic polarization planes, and we write

$$|\psi\rangle = \alpha \cdot |\updownarrow\rangle + \beta \cdot |\leftrightarrow\rangle, \tag{1.1}$$

where the values α and β are probability amplitudes, i.e., upon measuring an incoming photon by our detector, α is the chance for it to appear as horizontally polarized, and β is the chance for it to appear as vertically polarized. Alternative bases are also admissible, such as the *diagonal basis*, which describes the photon as a mix of $+45°$ polarization $|\nearrow\rangle$ and $-45°$ polarization $|\nwarrow\rangle$. Depending on whether a zero or one bit is transmitted, we write $|0\rangle$ or $|1\rangle$ to denote a respectively prepared qubit carrying the 0 or 1.

To read from the encoded information, we need to "measure" the photons, which means determining their polarization plane. However, the key point of using light

© Springer Nature Switzerland AG 2022
M. Mehic et al., *Quantum Key Distribution Networks*,
https://doi.org/10.1007/978-3-031-06608-5_1

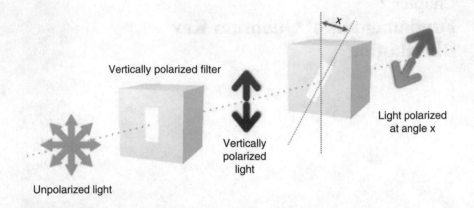

Fig. 1.1 Polarization of light

particles as carriers is their fragility: a qubit cannot be split, copied or amplified without introducing detectable disturbances, manifesting as measurement errors detectable by other means. This partly roots in Heisenberg's uncertainty principle [5] and is known as the no-cloning theorem [6]. It entails that each measurement inevitably modifies the quantum state so that an eavesdropper is unable to monitor communication without altering the transmitted information, which in turn can be detected by the sender and the receiver in the subsequent quantum protocol steps [4].

Let us analyze the example shown in Fig. 1.1, where non-polarized light enters a vertically aligned filter which absorbs some part of the light and polarizes the remainder in the vertical direction. The light is then directed through the second filter oriented at angle x to the vertical filter. A certain probability exists that the photon will also pass through the second filter. This probability depends on angle x. As x increases, the probability of the photon passing through this filter decreases until it reaches zero at $x = 90°$, when we consider this filter a horizontal polarizing filter. However, when $x = 45°$, a 50% probability exists that the photon will pass through the filter. The basis is defined as a pair of orthogonal polarization states which are used to describe the polarization of photons. The following bases can therefore be described:

- Rectilinear (vertical polarization, where $x = 0°$, and horizontal polarization, where $x = 90°$)
- Diagonal (diagonal polarization, where $x = 45°$ and $x = -45°$)

Consider communication between two users applying the polarization phenomenon to transmit bits. User A, typically called "Alice" in the literature, uses a filter in the rectilinear basis (either a horizontal or vertical polarizing filter) to polarize the photon and direct it to user B, typically named "Bob". Bob can reliably

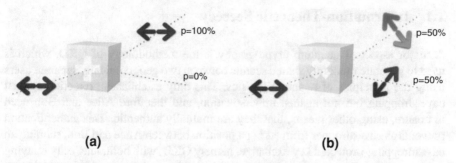

Fig. 1.2 Probability of detecting a horizontally polarized photon with (**a**) rectilinear or (**b**) diagonal basis

determine the polarization of the received photon only if he uses a filter aligned to the same, rectilinear basis (Fig. 1.2a). However, if Bob uses a filter in the diagonal basis, he cannot consider the received information reliable since the probability of receiving the photon is only 50%, which is not sufficient to accurately determine the basis used by Alice (Fig. 1.2b). Additional steps are therefore required where users exchange information about the sequences in which the same basis was used.

Correspondingly, according to the no-cloning quantum physical theorem, it is not possible to reproduce a photon with the same properties. More precisely, it is not possible to duplicate an unknown quantum state while keeping the original intact [6], therefore excluding a person-in-the-middle (PITM) attack. If an eavesdropper attempts to perform such an attack, he will be able to produce a photon with the same properties only with a certain probability, while other photons will not have the same properties.

Let us assume that the channel is being eavesdropped. If eavesdropper Eve uses the same basis as Alice, she can detect the original polarization of the photon. If, however, Eve uses a misaligned filter (diagonal basis), she will obtain no information (just as Bob in the previous case), and by doing so, change the initial polarization of the photon and Bob will receive a "garbled" photon. Finally, when Alice and Bob later exchange details of the basis they used via a public channel and disclose some of the obtained measurements, they will be able to detect the presence of Eve. Of course, an eavesdropper is not the only single possible cause of transmission errors. Measurement errors are also possible (e.g., by optical misalignment, noise in the detectors or disturbance of the quantum channel), but according to theory, such errors usually remain within known limits. If the error rate exceeds a "natural threshold", Alice and Bob can be certain that eavesdropping has occurred. It should be noted that quantum mechanics does not prevent the occurrence of eavesdropping, it enables its detection afterwards if it has occurred.

1.1 Information-Theoretic Secrecy

A major aspect of quantum cryptography is the methodology of QKD, which is used to generate randomly shared secrets between two geographically separate users using the principles of quantum physics. This only establishes protection against eavesdropping but not against impersonation, and therefore Alice and Bob need to ensure, using other means, that they are mutually authentic. This authentication prevents an eavesdropper from taking a position between Alice and Bob, running an eavesdropping-protected key exchange, namely QKD, with both, later only relaying the messages between Alice and Bob but invisibly reading all traffic as plaintext. This is a PITM attack, and its prevention calls for Alice and Bob to authenticate all messages between them, especially those related to the QKD protocol itself. For that purpose, the bit array generated by QKD cannot be used, since Alice and Bob need secrets for authentication in advance. Hence, given that Alice and Bob already need to start from a shared secret needed to authenticate the later QKD protocol steps, the more accurate term would be *quantum key growing*, describing that QKD cannot create keys out of nothing but relies on an initial authentication secret to generate a larger number of shared secrets between the peers. A typical use for such shared secrets is as encryption keys, but the purpose is not prescribed nor otherwise implied by the use of QKD. The keys are secure from eavesdropping during transmission, and QKD ensures that any third party's knowledge of the key is reduced to a minimum [7]. Other basic cryptographic primitives have also been raised to the quantum domain, such as quantum authentication, quantum secret sharing and quantum signature, but QKD, as of the time of writing this book, remains the major aspect of quantum cryptography.

The primary goal of QKD is to provide Information-Thoeretic Security (ITS) communication.[1] An information-theoretic secure system means that a system is still secure even if an attacker has unlimited resources available to perform a cryptographic analysis. This definition was presented by Claude Shannon in a well-known paper on "perfect secrecy" [8], which defined the perfect secrecy system as a system where the a posteriori probabilities of a ciphertext (encrypted message),

---

[1] One should not equate ITS with systems that come without computational intractability assumptions often referred to as *unconditionally secure*. Even QKD systems relying on the laws of quantum physics are based on some fundamental assumptions such as end-to-end confidentiality or the "trusted relay" concept which imposes the conditions for applied security. ITS expresses security in terms of entropy, mutual information, probability and generally without explicit reference to or assumption of any computational intractability; equivalently, assuming the adversary has infinite computational power. However, it *does* usually come with other assumptions, for example, the inability of the adversary to conquer certain nodes or a certain number of nodes (perhaps selected at random for an attack). *Unconditional security* is not a specific notion of security itself, as is ITS, but merely expresses the absence of certain assumptions, in most cases related to computational intractability. However, it does not mean "zero-assumption" security. The assumptions are simply different and may be partly implicit (as for QKD, the literature does not always name all the quantum laws behind its arguments, thereby its validity is an implicit, although unstated, assumption).

which are intercepted by an eavesdropper, are identical to a priori probabilities of the same message before interception. In other words, suppose that an attacker has some prior knowledge of which messages are likely and which are not. Then, obtaining the ciphertext, the attacker may be able to improve this prior guess, ideally up to the point of discovering the content which has been transmitted without any uncertainty. Shannon's definition of perfect secrecy then demands that this improvement should simply not occur, i.e., anything that the attacker knew about the potential secret content before having obtained the ciphertext should remain unchanged, even if the ciphertext is available for analysis. In more formal terms, if we write $P_c(m)$ for the probability of the secret content to be m if the ciphertext c is known, perfect secrecy requires $P_c(m) = P(m)$, where $P(m)$ is the (unconditional) probability for m to have been transmitted, which is what the adversary already knew before. Hence, the ciphertext would not reveal anything new to the attacker, regardless of what is done. Shannon showed that One-Time Pad (OTP) is an instance of such an unbreakable cryptosystem and provides perfect secrecy since it does not reveal any details of the original encrypted message (the uncertainty of the original message in terms of entropy remains unchanged if the ciphertext is known). A necessary and sufficient condition for perfect secrecy is obtained from Bayes' theorem, which states that perfect secrecy is given if and only if the chance for any particular ciphertext to come up by encryption is stochastically independent of the encrypted message m, i.e.,

$$P_m(c) = P(c), \text{ for all messages } m \text{ and all ciphertexts } c \qquad (1.2)$$

The OTP cipher was first presented in 1917 [9] by the American scientist Gilbert S. Vernam (AT&T Bell Labs), who introduced the most important key variant to the Vigenère cipher system. The "one-time" security requirement was discovered via an attack proposed by William Friedman, who described a method of gaining information, thus violating Shannon's definition (although by that time not yet existing), if the pad (key) is used twice or more often. Shannon later showed that for the OTP to provide perfect secrecy, the entropy of the key must necessarily be equal or greater than the entropy of the message to be encrypted. In case of the Vernam cipher, the sender performs an XOR-operation to encrypt the clear text message and secret key and creates a ciphertext, while the receiver uses the same operation to decrypt the ciphertext and obtains the clear text message.

The most significant problem of OTP is the secure distribution of such a large key between remote clients. In World War II, the one-time key material was printed on silk and concealed inside an agent's clothing. Whenever the key was used, it was torn off and burned [10]. Today, QKD can be used to exchange a secret key securely. The combination of QKD with OTP and an information-theoretic secure message authentication scheme such as Wegman-Carter [11, 12] is today typically carried under the term "quantum cryptography", although the field has grown considerably beyond key establishment only. In this narrower view, however, quantum cryptography enables ITS communication regardless of the advances made in mathematics or computer science, including quantum computation.

1.2 QKD Protocols

A QKD protocol defines procedures in the key material establishment process between remote users in a safe manner. Generally, three broad categories of QKD protocols can be identified: the oldest and widespread group of discrete-variable protocols (BB84, B92, E91, SARG04), more recent efficient continuous-variable (CV-QKD) protocols and distributed-phase-reference coding (COW, DPS) [13, 14]. The essential difference between protocols is reflected in the manner photons are prepared and transmitted over the *quantum channel* [4, 15–17]. Once a sufficient number of qubits has been exchanged, the protocol switches to a *classic channel*, e.g., the Internet, over which the qubit stream is compiled into the final key. These remaining steps, at a high level, are nearly identical to all protocols and often summarized as QKD *post-processing*. Figure 1.3 shows the general skeleton of a QKD protocol, regardless of implementation details, following a series of common steps: exchange of qubits (red arrow), extraction of the raw key (sifting), Quantum Bit Error Rate (QBER) estimation, key reconciliation, privacy amplification, and authentication [13] (all blue arrows). It is important to consider that the authentica-

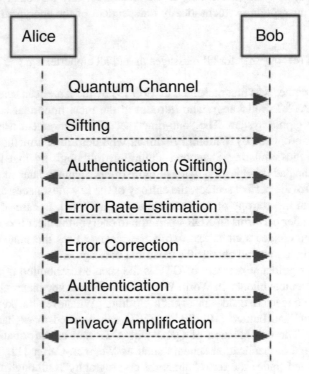

Fig. 1.3 QKD protocol—general key establishment procedure. Communication via the quantum channel (red line) is simplex, while post-processing phases are performed via the public channel in duplex mode (blue lines)

tion step is inherent to all other steps, and as such, not separately performed. Instead, all protocol messages exchanged over the classic channel must be authenticated.

1.2.1 BB84 Protocol

The Bennett-Brassard 1984 (BB84) is the oldest QKD protocol [3]. For the exchange of qubits over the quantum channel, it uses the two bases of orthogonal polarization, rectilinear and diagonal. Post-processing includes the following five steps performed over a public channel: sifting-extraction of the raw key, error rate estimation, error correction, privacy amplification, and authentication, with the latter again being inherent to all communication over the classical line.

Since BB84 defines the use of photon polarization as a means to carry information, Alice and Bob need to know the polarization state corresponding to each bit value. For example, bit value 1 is represented by either vertical or diagonal polarization, and bit value 0 is represented by horizontal polarization or opposite diagonal polarization. It is important to note that the two polarization states represent the same bit value. In the first phase of the key establishment procedure, Alice then generates random bits and polarizes the photons using according to the previously defined polarization states.

For example, for bit value 1, Alice can choose either diagonal polarization ($x = 45°$) or vertical polarization ($x = 90°$), and for bit value 0, either opposite diagonal polarization ($x = -45°$) or horizontal polarization ($x = 0°$). At the receiving end of the quantum channel, Bob uses a randomly selected basis for detection. Since Bob does not have information about the basis which was used to polarize the photon, he applies a randomly selected basis, which allows the reliable detection of only 50% of the sent key, as explained above (Table 1.1).

Now let us consider the situation when Eve attempts to listen on the line. Since Eve does not know which basis Alice has used, she needs to use a random basis to detect the photon's polarization. If her measurement basis is correct, it will not change the polarization of the photons; if she uses the incorrect basis, the original polarization of the photon will change (Table 1.2), this change being likely detectable.

For the first bit, Eve uses the correct basis for detection and obtains the correct information without changing the initial polarization of the photon. However, for the second bit, Eve uses the incorrect basis for detection. This changes the original polarization. As a result, Bob will also incorrectly measure, even if he uses the correct basis for detection. Although he does not yet know that Eve was listening, the error was irreversibly made, and Bob can later discover it. The resulting material obtained through the quantum channel is often denoted as raw key.

Table 1.1 An example of key distribution in the BB84 protocol

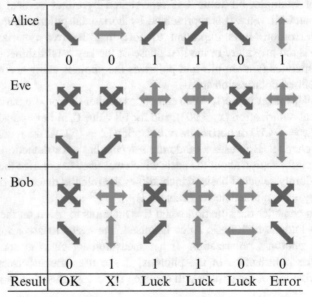

Alice	↘	←	↗	↗	↘	↑
	0	0	1	1	0	1
Bob	✕	✚	✕	✚	✚	✕
	↘	←	↗	↑	↑	↘
	0	0	1	1	1	0
Result	OK	OK	OK	Luck	Error	Error

Table 1.2 An example of eavesdropping in BB84 protocol

Alice	↘	←	↗	↗	↘	↑
	0	0	1	1	0	1
Eve	✕	✕	✚	✚	✕	✚
	↘	↗	↑	↑	↘	↑
	0	1	1	1	0	0
Bob	✕	✚	✕	✚	✚	✕
	↘	↑	↗	↑	←	↘
	0	1	1	1	0	0
Result	OK	X!	Luck	Luck	Luck	Error

1.2.1.1 Sifting: Extraction of a Raw Key

Once the exchange of qubits over a quantum channel has been accomplished, Alice and Bob continue on the classic channel. First, they engage in a conversation to compare the bases they have used to send or measure qubits. To this end, Bob informs Alice of the bases he used, and Alice provides feedback advising when incompatible bases were used. It is important to highlight that information concerning the measurement results is not revealed, as only the bases are publicly revealed, and Eve may listen on this conversation without problems or being detected. However, authentication should prevent Eve from actively manipulating the exchanged information, which is again critical for security. Upon receiving Alice's information where Bob was correct or mistaken regarding the measurement bases, Bob discards all the results where he applied the incorrect polarization filter and keeps only those qubits measured with the correct polarization filter. Therefore, the length of the raw key is further reduced and the key material available for further processing after sifting phase is often referred to as the sifted key.

1.2.1.2   Error Rate Estimation

Alice and Bob then check whether the error rate was below what would be naturally expected or exceeds the known limit which indicates Eve's presence. By taking measurements, Eve necessarily adds something to the natural error rate p_{max}. This value is public knowledge on the quantum channel without the presence of Eve and establishes the standard for the QBER p which Alice and Bob will determine through their public discussion. The usual approach for estimating the QBER p is to compare a small random sample portion of measured values. The selected portion should be sufficient to make the estimated QBER credible, while not sacrificing too much of the raw key (remember that Eve obtains all information as plaintext over the public channel).[2] The choice of how many bits to uncover (also to Eve) is a matter of a cost-benefit trade-off: using too few makes the QBER approximation unusefully inaccurate, while using too many for an accurate QBER estimate will make the overall resulting key perhaps too short, and the overall QKD becomes inefficient. Niemec [18] remarked that comparing between 8 and 37% are sufficient for collecting information about the security of a distributed key, and hence suitable for personal and ordinary commercial use of quantum cryptography.

The estimated QBER value p can be compared with the already known threshold value p_{max}. If the error rate is greater than a given threshold ($p > p_{max}$), the presence of Eve is revealed, meaning that all measured values should be discarded and the process should start from the beginning. Kollmitzer recommended that the threshold value be set to 11% [4], while a maximum tolerated error rate of 12.9% was suggested in [19]. If, however, the estimated QBER is lower than the threshold value p, eavesdropping is considered not to have occurred, and the next steps of the protocol can be taken.

---

[2] It is natural to propose that Alice and Bob can simply encrypt their public discussion to make it more difficult for Eve to gain information. However, Alice and Bob cannot do this since (i) they explicitly want to avoid public-key, i.e., encryption based on intractability, and (ii) they are not yet in possession of a key sufficiently long to protect their entire public discussion. It also follows from the results of Shannon that no key of fixed length can ever provide perfect secrecy, therefore the initial authentication secret which Alice and Bob possess does not lend itself to an ITS encryption of their public discussion.

1.2.1.3 Error Key Reconciliation

The result of the QKD process is the establishment of a key for symmetric cryptography between remote entities. This requires that no differences exist between the keys which Alice and Bob finally obtain, otherwise, the key is useless.

Even if Alice and Bob determine Eve's absence with a low empirical QBER estimate, they will nonetheless need to correct the residual lot of errors caused by natural effects such as the imperfections of optical media, sources, signal detectors or other phenomena in photon transport and measurement. In this chapter, the terms "detection" and "discardance" or errors are used synonymously. Although minor differences exist, the error key reconciliation process attempts to detect the positions of errors in the key, which can then be treated easily: subsequent operations may include overwriting the identified error positions to default values (i.e., setting all bits on detected positions to the default value "0") or simply discarding bits on detected positions from further use. The process of locating and removing errors is often denoted "error key reconciliation". The main aim is to locate errors while leaking only a minimum amount of information about the key to Eve. Additionally, the solution must be efficient in computation and communication, given that time-consuming or exhaustive algorithms can lead to downtime of the public and quantum channels (more details in Chap. 4).

Cascade

One of the oldest and simplest approaches is a protocol called Cascade [20]. It was reported by Brassard and Salvail in 1994 as an improved version of the BBBSS protocol, which was developed in 1991 for the purposes of the first practical QKD channel experiment. The protocol works most effectively where no consecutive errors (burst) are present. It begins by applying a random permutation on the sifted key to evenly disperse the errors. The permuted sifted key is then divided into blocks of size k, where size k is selected such that each block should contains no more than one error. Defined in this manner, block size k depends on the estimated QBER value p and is an essential parameter of the Cascade settings. The authors analyzed a suitable value, and based on empirical analysis, proposed $k = 0.73/p$, where p is the empirically determined error rate (fraction) [20]. Alice and Bob exchange parity values for each of the defined blocks. If the parity values are identical, these blocks are no longer considered. However, if the values are different, a detailed binary search is performed until the error which causes the different parity values is detected. The binary search implies a further division of the blocks into smaller sub-blocks for which the parity values are exchanged until an error is detected.

At this stage, Alice and Bob are certain that no odd number of errors exists in their blocks. As shown in Fig. 1.4, more than 50% of errors will be identified after the first iteration, and the detection success rate increases linearly with the QBER value. However, it is possible that an even number of errors will "mask" the parity

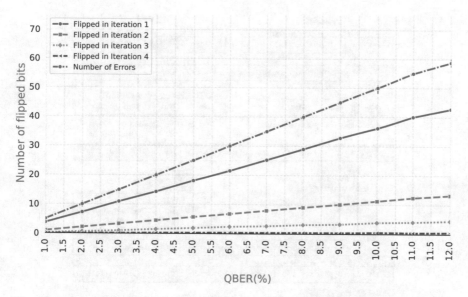

Fig. 1.4 The average success rate of Cascade iterations versus QBER. The length of the sifted key is $n = 1000$ bits. Each dot marks the average value calculated from 1000 different simulations per QBER value

value. It is therefore necessary to perform an additional iteration by applying a new permutation of the sifted key and exchanging the parity values again for the defined blocks. To reduce the communication load created by exchanging a large number of small packets containing only parity values, the block size is doubled $k_i = 2 \cdot k_{i-1}$ in each subsequent iteration i. Two passes of the Cascade are illustrated in Fig. 1.5.

Cascade is known as a communication-intensive protocol because it is necessary to exchange one bit for each parity value. Note that the parity values of all the blocks can be placed into a single packet. However, in the binary search process, Alice and Bob will have to exchange separate packets with each calculated parity value, up to a maximum $1 + \lceil \log_2 k_i \rceil$ packets, leading to a large communication overhead. Additionally, because all communication takes place on the public channel, there is a risk that Eve will discover information about the sifted key. The last bit of the block or sub-block in the binary search process for which the parity value was exchanged should be discarded, as proposed for BBBSS [21]. The maximum number D of revealed bits can be calculated from k_i, as follows:

$$D = \sum D_i = \sum_i \left(\sum_{\substack{initially \\ even \\ blocks}} 1 + \sum_{\substack{initially \\ odd \\ blocks}} (1 + \lceil \log_2 k_i \rceil) + \sum_{\substack{other \\ errors \\ corrected}} \lceil \log_2 k_i \rceil \right)$$

(1.3)

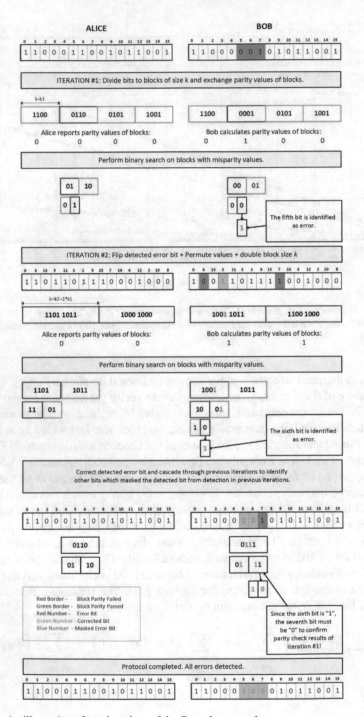

Fig. 1.5 An illustration of two iterations of the Cascade protocol

This can be shortened to:

$$D = \sum_i D_i = \sum_i \left(\frac{n}{k_i} + \sum_{\substack{errors \\ corrected}} \lceil \log_2 k_i \rceil \right) \tag{1.4}$$

where $k_i = 2 \cdot k_{i-1}$, $k_i < \frac{n}{2}$ and n denote the length of sifted key [22].

Although 20 iterations are planned for the BBBSS protocol, analyses of the operation of the Cascade demonstrated that four iterations were sufficient [23]. Since the value of k is doubled for each iteration, it is recommended that the protocol be terminated when the value of k is equal to or greater than the length of the sifted key. Practical experiments show that in most cases (especially for smaller QBER values) a maximum of three iterations is sufficient. It is important to point out that these approaches apply adaptive block sizes k. This type of modification has proven to be effective and is discussed in more detail in Chap. 4.

The process described above is identical for both the BBSS and Cascade protocols, with one significant difference. During the detection of errors in the second and subsequent iterations, the Cascade protocol concludes that an even number of errors must have been in the blocks of the previous iterations. The detected error and the one which "masked" the parity value of the block can be detected by performing a binary search of the block from the previous iterations. Therefore, in the Cascade protocol, the information leakage reduction based on discarding the last bit of each subblock cannot be performed as it would disable the recursive process. Instead, the Cascade uses the number of revealed bits denoted by the number of exchanged parity blocks to reduce Eve's knowledge of the key in the subsequent privacy-amplification step.

Figure 1.5 illustrates that this means the exchange of information concerning a block with different parity values in subsequent iterations will reveal an even number of errors. This recursive feedback enables a significant improvement of the protocol, after which the Cascade protocol is named.

The Cascade protocol has been studied extensively, and the practical implications for other phases of post-processing have also been investigated [24, 25]. An empirical analysis of the Cascade protocol was presented in [26]. Use of interleaving, which entails splitting bursts of errors into smaller chunks, was suggested in [27]. A broad analysis of the initial value of parameter k was reported in [28, 29].

Winnow Protocol

In the Cascade protocol, the binary search dominates the communication process. An alternative which reduced this intensity, called the Winnow protocol, was proposed in 2003 [30]. The Winnow protocol is based on Hamming codes and establishes that the use of generator matrix G and parity check matrix H are defined such that $H \cdot G^T = 0$. The rate of the code is given by $r = \frac{m}{n}$, where m and n are the dimensions of the generator matrix G. To illustrate our case, we use a (7,4) Hamming code which embodies 4 information bits in a 7 bit code word.

Sender calculates: $M \cdot G = C$

$$[1\ 0\ 1\ 0] \cdot \begin{bmatrix} 1\ 0\ 0\ 0\ 0\ 1\ 1 \\ 0\ 1\ 0\ 0\ 1\ 0\ 1 \\ 0\ 0\ 1\ 0\ 1\ 1\ 0 \\ 0\ 0\ 0\ 1\ 1\ 1\ 1 \end{bmatrix} = \begin{bmatrix} 1 \\ 0 \\ 1 \\ 0 \\ 1 \\ 0 \\ 1 \end{bmatrix}$$

Receiver calculates: $C \cdot H = S$

$$\begin{bmatrix} 1 \\ 0 \\ 1 \\ 0 \\ 1 \\ 0 \\ 1 \end{bmatrix} \cdot \begin{bmatrix} 0\ 1\ 1\ 1\ 1\ 0\ 0 \\ 1\ 0\ 1\ 1\ 0\ 1\ 0 \\ 1\ 1\ 0\ 1\ 0\ 0\ 1 \end{bmatrix} = [0\ 0\ 0]$$

Using Hamming code, the sender calculates the dot-product of the message M and generator matrix G to produce the *code word* which is delivered to the receiver ($M \cdot G = C$). The receiver then computes the dot product of the code word and the parity check matrix H to produce the *syndrome* ($C \cdot H = S$). If the syndrome is a zero-vector, the receiver concludes no errors in the current messages, otherwise at least one error is present for potential detection by the syndrome. The most usually applied method is known as maximum likelihood decoding, which establishes that the nearest code word is the correct one. The number of positions between two code words is known as the Hamming distance d_{min} and limits the protocol's efficiency. If the number of errors between messages is less than $d_{min}/2$, Hamming code can detect the error since the nearest code word is the one which is correctly detected.

This method of using Hamming codes is known as Forwarding Equivalence Class (FEC), which sends redundant bits along with the message (note that $C > M$). For application with QKD, modification is necessary so that the message is not transmitted as a result of eavesdropping. Alice and Bob therefore divide their sifted keys M_a and M_b into equal blocks of size $k = 8$. For each block, Alice calculates the dot-product to produce the syndrome $S_a = H \cdot M_a^T$, which is sent to Bob. Bob also calculates the dot-product to produce his syndrome $S_b = H \cdot M_b^T$ for each block. Exchange of the calculated syndrome is similar to the exchange of parity values of blocks and sub-blocks when performing binary search in Cascade.

$$S_a = H \cdot M_a^T = \begin{bmatrix} 0 1 1 0 1 0 1 \\ 1 0 1 1 0 1 0 \\ 1 1 0 0 1 0 1 \end{bmatrix} \cdot \begin{bmatrix} 0 1 1 0 0 1 1 \end{bmatrix}^T = \begin{bmatrix} 1 \\ 0 \\ 0 \end{bmatrix}$$

$$S_b = H \cdot M_b^T = \begin{bmatrix} 0 1 1 0 1 0 1 \\ 1 0 1 1 0 1 0 \\ 1 1 0 0 1 0 1 \end{bmatrix} \cdot \begin{bmatrix} 0 1 0 0 0 1 1 \end{bmatrix}^T = \begin{bmatrix} 0 \\ 1 \\ 0 \end{bmatrix}$$

$$S_d = S_a \oplus S_b = \begin{bmatrix} 1 \\ 1 \\ 0 \end{bmatrix}$$

$$1 \cdot 2^0 + 1 \cdot 2^1 + 0 \cdot 2^2 = 3$$

The error is on the 3rd position.

The Winnow protocol is able to detect only an odd number of errors. Additional iterations similar to Cascade must therefore be performed and the entire process repeated. The size of the block k increases with each regeneration of the parity check matrix. Although this significantly improves performance in error key reconciliation, the Winnow protocol introduces an additional problem: because of its dependence on Hamming codes, it may introduce errors if the number of errors per block is too high. If the number of errors is greater then the Code's Hamming distance d_{min}, decoding would give Bob a string which is different to Alice's result and thus render the key unusable. To reduce an eavesdropper's knowledge of the key and speed up work performance, preliminary paired block comparison values are exchanged. If the parities match, it means that there is an even number of errors that will be detected in subsequent iterations after performing random permutations. But, if the parities do not match, there is an odd number of errors in the block. When the number of errors is sufficiently small and randomly distributed throughout the data, the parity mismatch indicates a single error per block. Thus, in those cases, syndrome values are exchanged to detect the error bit. Unlike the Cascade protocol in which a subsequent privacy amplification phase is performed, in Winnow, the *privacy maintenance* is performed throughout the reconciliation phase. One bit is discarded for each of the exchanged parities, provided that the bit selection is not random. The authors claim that this approach is more efficient because some of bits discarded may be bits in error. When the size of blocks is defined as $k = 2^i$, $i = 3, 4, \ldots$, a single bit is discarded due to parity comparisons reducing the block size of $k = 2^i - 1$ before computing syndrome. Thus, the amount of bits discarded per odd block is:

$$N_{discarded}^{oddblock} = \log_2(k) + 1 \tag{1.5}$$

The Winnow protocol requires a reliable initial assumption on error distribution and how many errors should be expected. Compared to Cascade, Winnow reduces communication complexity because up to two messages are exchanged per block (one about parity values and the other with syndrome values).

Low Density Parity Check

Although Cascade and Winnow are proven protocols which have been widely used in practical QKD systems, their communication complexity can lead to reduced performance. Namely, both protocols require the exchange of parity/syndrome values for each blocks and sub-block. Consider a sifted key of length 1000 bits with a QBER of 6%, or more precisely, a total of 60 error bits. Cascade will form $k_1 = 0.73/0.06 = 77$ blocks of 13 bits and flip approximately 46 bits in the first iteration, 12 bits in the second and 6 bits in the third. Considering that block parity values can be transmitted in a single packet and sub-block parity values in separate packets, around 200 packets must be exchanged to detect all errors. The Winnow protocol significantly reduces communication because it excludes binary search, but it may leave undetected errors.

The main disadvantage of these protocols is their limitation in being able to correct only one error per block. This leads to the use of bit shuffling in multiple iterations, thus requiring additional signaling communication, and is the reason why current commercial QKD systems generally turn to more efficient solutions such as Gallager's Low Density Parity Check (LDPC) [31–33]. LDPC codes are known for the scarcity (low density) of the parity check matrix, which allows them to grow almost linearly in decoding complexity with respect to message length. Another alternative is to make the block size dynamic w.r.t. a known (or assumed) error scattering pattern, so that only one error per block can be expected. This *adaptive* version of Cascade, has its own Sect. 4.1.1.1 dedicated to it.

Similar to Hamming codes, LDPCs are used as FEC codes and use a parity matrix H and generator matrix G. The dimensions of matrices $m \times n$ are defined using the code rate r, which lies in the range [0,1] such that $m = n(1 - n)$. The parity check matrix is mainly populated with zeros, where the number of ones (which define the number of parity checks) grows linearly with increases in n. The width of H shrinks with an increase in r, while H grows quadratically with increases in n. The r value defines the efficiency of the protocol, and it is usually defined in advance. This definition of LDPC code is known as regular LDPC code. Subsequently, extensions where the number of ones in each column and row is not fixed and depends on the degree of distribution are denoted irregular LDPC codes.

The application of LDPC codes in QKD is similar to the application of Hamming code used in the Winnow protocol. The difference is that instead of generating a symbol for one block, a symbol is generated for the entire sifted key. The steps in executing the LDPC protocol are as follows:

- Alice and Bob estimate the QBER.
- Based on the estimated QBER, they will choose a parity check matrix H and generator matrix G.
- Alice calculates syndrome S_a and sends it to Bob.
- Based on Bob's key, the received syndrome S_a, the parity-check matrix H, and estimated QBER value, Bob attempts to reconcile the sifted key which is attempting to resolve Alice's key. Several decoding techniques such as the iterative belief propagation decoding algorithm (also known as the Sum–Product algorithm) or log-likelihood ratios significantly lower the computational complexity of this process [4, 25, 34].
- Alice and Bob discard the number of bits defined by size m of the exchanged syndrome.

It is known that decoding LDPC code is a computationally expensive and memory intensive solution. Consider a key of 1000 bits and rate code of 0.1. The parity check matrices of LDPC should be $900 \times 1000 = 9 \times 10^5$ bits. If keys of larger size are considered, for example 100k bits, it will include 9×10^9 bits. The migration of error key reconciliation solutions to FPGA boards with the aim of speeding up the operation are therefore noticeable [35–38]. However, the reduction in communication load represents a significant advantage over the Cascade and Winnow protocols. LDPC has been applied in a number of QKD projects, for example, in developing the BBN Niagara approach in the DARPA QKD network [39], for post-processing in the Toyko QKD network [40], and others. The use of artificial neural networks for error reconciliation was proposed by Niemiec [41].

Comparison

This section compares the performance of the Cascade, Winnow and LDPC protocols. The program code was implemented in C++ and installed on a server with an Intel(R) Xeon(R) Silver 4116 CPU @ 2.10 GHz, 8 GB RAM and a 512 GB HDD. For each protocol and each QBER value, 10,000 simulations were performed with different seed values which determined the error positions. A total of 870,000 simulations were thus performed. The code was implemented on a single server so that Alice and Bob applications were executed locally on the same machine. The protocols were operated under optimal network conditions, i.e., no network delay which could slow down the protocol operations was present.

The results in Fig. 1.6 show the number of exposed bits which should be discarded to reduce Eve's knowledge of the sifted keys obtained during the error reconciliation phase. The number of exposed bits is defined as follows:

- Cascade: One bit is discarded for each parity value exchange.
- Winnow: One bit is discarded per block k.
- LDPC: The number of discarded bits is defined according to the length of the exchanged syndrome S_a.

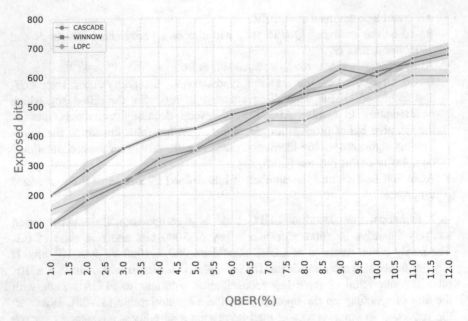

Fig. 1.6 Comparison of the number of information bits exposed for different values of QBER. The number of information bits exposed for LDPC is the length of the syndrome exchanged; for Cascade it is the number of exchanged parity values; for Winnow it is the number of exchanged parity values and the length of syndromes exchanged. The length of the sifted key is $n = 1000$ bits

We can conclude that for small QBER values (up to 5%), Cascade has good performance. If few errors are present, the number of binary searches is reduced. However, as the QBER increases, Cascade requires more detailed searches, resulting in a greater number of rejected bits. LDPC demonstrates better efficiency in terms of discarded bits, as only one syndrome is transmitted.

The results in Fig. 1.7 show that the complexity of the LDPC code to generate only one syndrome based on the parity check matrix requires significantly more time than Cascade and Winnow protocols. Given that protocols were executed without network delays, it is clear that Cascade and Winnow have almost fixed execution times.

1.2.1.4 Privacy Amplification

To reduce the amount of information potentially obtained by Eve, further reduction of the key is recommended through a step called privacy amplification. The step involves the rejection of some bits, whose number can be calculated from Eq. (1.6), where D is the number of bits which will be discarded and n is the length of the reconciled key [42].

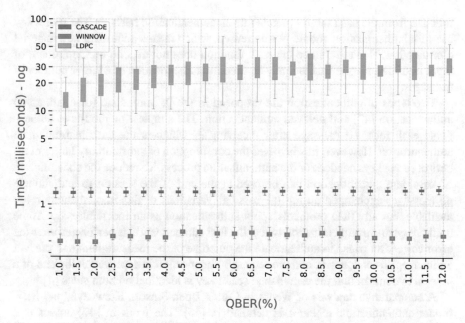

Fig. 1.7 Comparison of the execution time on local machines for different values of QBER

$$\frac{n \cdot 2^{-D}}{\log 2} < 1 \tag{1.6}$$

We refer interested readers to find more information concerning entropy measurement in a detailed work by Cachin [43]. It is important to mention that a quantum channel only permits simplex (one-way) communication, even though Alice and Bob can send messages in both directions over a classic channel [16]. The classic channel does not rely on the laws of quantum physics, therefore Eve is able to listen without penalty to all communication on this channel. To prevent a PITM attack, situations where an eavesdropper is able to alter the messages being sent require authentication of the classic channel.

1.2.1.5 Authentication

The authentication problem was described in the original BB84 paper [3], in which Bennet and Brassard proposed a solution based on universal families of hash functions introduced by Wegman and Carter [44]. Although more effective symmetric authentication methods are known today, the Wegman–Carter authentication is often used since it depicts an upper bound for the required symmetric authentication key [12, 45, 46]. Secure authentication of the public channel requires both communication parties to pre-share a small secret key. Alice and Bob must therefore possess a small portion of the secret key which can be used to select a hash function from the hash family to generate an authentication hashtag. The secret key used for authentication should not be reused, which means additional key material consumption. However, since the hash function does not depend on the amount of input information, one hashtag may be used for authentication of a larger number of messages.

Two types of authentication are proposed in QKD: immediate (or continuous) authentication [47] and delayed authentication. The simplest method is to authenticate each received message after receiving it, which is the case in immediate authentication. However, this involves the consumption of substantially larger numbers of secret keys needed for the authentication process. To reduce the consumption of keys, late authentication can be used at the end of the session to authenticate all messages exchanged during the session. Variations of this implementation are available, but all QKD protocols apply authentication as an inevitable step. Some methods perform authentication twice: the first authentication is performed after the error correction phase to authenticate the outcome of the measurement and prevent an intercept/resend attack, and the second authentication is deployed at the end of a session to confirm that the established secret key is identical on both sides [4].

A comparative analysis of Wegman–Carter, Boer, Sinson, Krawczyk, and Bierbrauer authentication schemes is detailed in [45]. The work in [47] argues that Wegman–Carter authentication based on ASU_2 (Almost Strongly 2-Universal) hashing is suitable for QKD authentication. As with any universal hashing scheme, the actual "hash function" comes as an entire *set* of functions $\mathscr{H} = \{H_1, H_2, \ldots\}$, in which each H_i is another fixed mapping into the set of hash-tags. The actual hashing is accomplished by randomly selecting an element from $\mathscr{H}$ and using the authentication secret to execute exactly this random choice. Wegman and Carter have proposed suitable classes of hash functions and also quantified the size of the index k to select a member from $\mathscr{H}$, which in our case is the authentication secret, according to:

$$k(g, c) = 4 \cdot (g + \log_2 \log_2 c) \cdot \log_2 c \qquad (1.7)$$

where c is the length of the messages to be authenticated and g is the length of the authentication tag, both in bits.

In delayed authentication, one authentication message should be exchanged to verify measurement on the quantum channel.

1.2.2 B92 Protocol

Bennett introduced the B92 QKD protocol, which relies solely on two non-distinguishable quantum states [48]. As with the BB84 protocol, Alice will generate a random binary sequence, for example, $S_A = [10110101\ldots]$ of length Q, and instead of using bases to modulate photons, she can use non-orthogonal polarizations such as:

$$|\varphi\rangle = \begin{cases} |0\rangle & \text{if } S_A[i] = 0 \\ \frac{|0\rangle + |1\rangle}{\sqrt{2}} & \text{if } S_A[i] = 1 \end{cases} \tag{1.8}$$

where i is the ith bit of the sequence S_A. The modulated photons are transmitted over a quantum channel where Bob performs measurements. Bob generates his own random sequence, for example, $S_B = [00101010\ldots]$ since no information about the sequence S_A is known to him at this point. As with the BB84 protocol, Bob applies rectilinear or diagonal polarisation bases to perform measurements. When the correct basis is used, the incoming photon is correctly measured. However, the measurement with the incorrect basis will not provide a result; this is often defined as "erasure" in quantum mechanics [49]. After the measurements, Bob will inform Alice in which positions the measurement has yielded results. The positions of bits in sequences S_A and S_B for which erasure measurements occurred will be discarded, while no information concerning polarization is revealed. Alice and Bob will discard bits which correspond to the photons Bob measured with an incorrect basis, while no information concerning the used basis is revealed to Eve.

B92 consists of the same post-processing stages (sifting-extraction of the raw key, error rate estimation, error correction, privacy amplification and authentication) as previously explained for the BB84 protocol [50].

1.2.3   CV-QKD

Continuous-Variable QKD (CV-QKD) is a special group of QKD protocols in which information about the key is not carried by a single photon but transmitted using the wave properties of light and employing multi-photon quantum states [51, 52]. The use of optical homodyne detection [53] means that CV-QKD is a cost-effective solution requiring only modest technological resources, although it raises security issues.

In the late 1990s, a new proposal to use squeezed states for QKD appeared [54]. This solution assumed that sequences of symbols are impressed on two squeezed beams mixed in a 50:50 beam splitter by Alice. Using homodyne detection, Bob can recover one of the two sequences. Hillery [55] later proposed sending displaced squeezed vacuum states which are squeezed into one of two orthogonal field quadrature components. Bob can then randomly choose which of the components to measure. However, these two solutions assumed that a randomly selected part of the sequence is used only to establish security through a public channel. A solution with greater efficiency [56] could use Einstein-Podolsky-Rosen correlations [57] between two beams to check whether eavesdropping has occurred. Cerf et al. [58] then proposed an alternative squeezed state QKD scheme in which both the key and quantum variable carrying it are continuous. This Gaussian-modulated squeezed state protocol allows Alice and Bob to share a key which contains a random list of Gaussian-distributed variables unknown to the eavesdropper.

Grosshans and Grangier [59] proposed the coherent state balanced homodyne detection protocol. This solution, now known as the GG02 protocol, assumes that Alice randomly modulates a Gaussian beam (both phase and amplitude are modulated with Gaussian random numbers) and sends it to Bob, who measures either the phase or the amplitude of the beam and informs Alice which measurement he made. Alice and Bob then have two correlated sets of Gaussian variables from which they can extract a secret key. Later a new coherent state QKD protocol was proposed [60]. The protocol eliminated the need to randomly switch between measurement bases and increased key rate by using simultaneous quadrature measurements. Practical implementation soon also confirmed the usefulness of these CV-QKD protocols [61, 62].

A new approach to CV-QKD based on multi-way quantum communication combines two physical resources in two-way quantum communication in which the first resource is used to assist the second. This approach increased the security of the key [63]. Even though the security of CV-QKD protocols has been widely analyzed and theoretically proved [64–69], a serious threat remains in the form of side-channel attacks which exploit detection using a publicly shared high-power local oscillator (LO). The security of QKD based on continuous variables can be compromised if an eavesdropper is able to manipulate the LO, which is propagated through the insecure quantum channel [70, 71]. An eavesdropper who has access to both the quantum signal and the LO can perform sophisticated attacks by manipulating the LO [72]. However, the risk of these attacks can be reduced by solutions which assume coherent detection using a "more locally" generated LO [73].

The first long-distance transmission using CV-QKD generated secret keys at a rate of approximately 2 kb/s over 25 km of optical fiber [74]. This implementation of a reverse-reconciliation coherent-state CV-QKD system was secured against collective attacks and supported authentication, reverse reconciliation and privacy amplification processes. In 2013, the practical generation of a secret key over the same distance increased to 10 kb/s using the GG02 protocol [75]. The maximum transmission distance for this implementation was 80.5 km (the system generated several hundred secret bits per second over this distance). A few years later, an experiment using the Gaussian-modulated coherent states protocol confirmed that practical implementation of CV-QKD over a distance of 100 km was possible [76] with a key rate of up to 1 kb/s. However, CV-QKD protocols based on a discrete modulation allow longer distances to be achieved than those based on Gaussian modulation. The most recent works therefore suggest that the possible distance for a discretely-modulated four-state CV-QKD protocol [77] is much higher than currently achieved, the expected maximum transmission distance for these long-distance protocols being up to 300 km [78].

1.3 Key Length

As discussed in the previous sections, the length of the final key depends mainly on the quantum bit error rate in the quantum channel. Additionally, it depends on the techniques for error estimation and error reconciliation which affect the numbers of bits discarded to reduce Eve's overall knowledge of the secret key. These techniques also have a significant effect on communication through the public channel [24]. Moreover, it is important to emphasise the importance of key management. A longer key can be reduced when needed, while smaller keys may be combined to form a longer key. To perform these and similar operations correctly, a key management system (key block identification, key merging, key splitting, deletion of used keys, etc.) is necessary.

The average amount of key material which can be established per unit of time, often referred to as the *key generation rate* or simply *key rate*[3] may vary with temperature, the stability of the quantum devices, humidity, pressure, global radiation, duration of sunshine, dust or other factors [4, 16]. However, the most significant impact is from the length of the link (Fig. 1.8). The key rate is an essential aspect of further research since it determines the type of encryption which can be used and the type of traffic which can be secured. Tight finite-key analysis for QKD protocols is reported in [80]. More theoretical and practical details of QKD protocols can be found in [13, 16, 47, 81, 82].

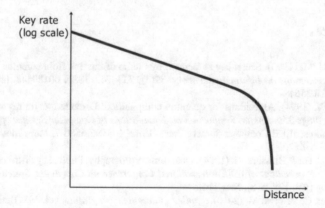

Fig. 1.8 Illustration of the dependence of key generation rate vs distance. The linearity of the curve is disturbed significantly because of increased dark counts in the detector and the presence of Raman noise. To increase the linearity of the curve, the detector can be placed in colder areas ($-60\,°C$) [83]

---

[3] Most of the terms used in this book follow the vocabulary defined with the ETSI 007 standard [79].

1.4   Summary

Quantum cryptography is an attractive cryptographic technology which has received the attention of various organizations among academic and industrial communities. In recent years, notable progress in the development of optical equipment has been reflected through a number of successful demonstrations of QKD technology. These demonstrations show great achievements in quantum cryptography and highlight the practical difficulties which still need to be resolved. Quantum routers and quantum repeaters are necessary for extending the secure transmission distance of quantum channels. Solutions to integrate QKD networks into existing optical communications networks are currently the major topic in optical research. In industry, the standards for the evaluation of security and the production and application of QKD have already been defined [84].

In addition to considering the efforts underway to facilitate data and quantum communication through the same optical link, it is important to note that the public channel between nodes is not limited to quantum traffic. A public connection in these cases is shared with other network traffic. For high-load business applications, QKD networks must guarantee QoS and be able to differentiate traffic according to the importance of the message being transmitted. To satisfy these requirements, the network must have suitable QoS solutions which can successfully resolve the race between key generation and key consumption.

References

1. Maurer, U. M. (1993). Secret key agreement by public discussion from common information. *IEEE Transactions on Information Theory, 39*(3), 733–742. ISSN 00189448. https://doi.org/10.1109/18.256484
2. Shor, P. W. (1994). Algorithms for quantum computation: Discrete logarithms and factoring. In *Proceedings 35th Annual Symposium on Foundations of Computer Science* (pp. 124–134). Los Alamitos: IEEE Computer Society Press. ISBN 0-8186-6580-7. https://doi.org/10.1109/SFCS.1994.365700
3. Bennett, C. H., & Brassard, G. (1984). Quantum cryptography: Public key distribution and coin tossing. In *Proceedings of IEEE International Conference on Computers, Systems and Signal Processing* (vol. 175, p. 8). New York.
4. Kollmitzer, C., & Pivk, M. (2010). *Applied quantum cryptography*, vol. 797. Berlin: Springer. ISBN 364-2-04829-3. https://doi.org/10.1007/978-3-642-04831-9
5. Heisenbergm, W. (1927). Über den anschaulichen Inhalt der quantentheoretischen Kinematik und Mechanik. *Zeitschrift für Physik, 43*(3–4), 172–198. ISSN 1434-6001. https://doi.org/10.1007/BF01397280
6. Wootters, W. K., & Zurek, W. H. (1982). A single quantum cannot be cloned. *Nature, 299*(5886), 802–803 (1982). ISSN 0028-0836. https://doi.org/10.1038/299802a0
7. Renner, R. (2005). *Security of Quantum Key Distribution*. Ph.D. Thesis, Swiss Federal Institute of Technology Zurich. http://arxiv.org/abs/quant-ph/0512258
8. Shannon, C. E. E. (1984). Communication theory of secrecy systems. *Bell System Technical Journal, 15*, 57–64. ISSN 0724-6811

9. Vernam, G. S. (1926). Cipher printing telegraph systems: For secret wire and radio telegraphic communications. *Transactions of the American Institute of Electrical Engineers, 45*(2), 109–115. ISSN 0095-9804. https://doi.org/10.1109/JAIEE.1926.6534724

10. Marks, L. (2001). *Between silk and cyanide: A Codemaker's war, 1941–1945*. New York: Simon and Schuster. ISBN 074-3-20089-6

11. Abidin, A., & Larsson, J.-Å. (2011). Security of Authentication with a fixed key in quantum key distribution. p. 14. http://arxiv.org/abs/1109.5168

12. Portmann, C. (2014). Key recycling in authentication. *IEEE Transactions on Information Theory, 60*(7), 4383–4396. ISSN 00189448. https://doi.org/10.1109/TIT.2014.2317312

13. Scarani, V., Bechmann-Pasquinucci, H., Cerf, N. J., Dušek, M., Lütkenhaus, N., Peev, M. (2009). The security of practical quantum key distribution. *Reviews of Modern Physics, 81*(3), 1301–1350. ISSN 0034-6861. https://doi.org/10.1103/RevModPhys.81.1301

14. Van Assche, G. (2006). *Quantum cryptography and secret-key distribution*. Cambridge: Cambridge University Press. ISBN 978-0-521-86485-5

15. Dodson, D., Fujiwara, M., Grangier, P., Hayashi, M., Imafuku, Kentaro, K.K., Kumar, P., Kurtsiefer, C., Lenhart, G., Luetkenhaus, N., Matsumoto, T., Munro, W. J., Nishioka, Peev, T. M., Sasaki, M., Sata, Y., Takada, A., Takeoka, M., Tamaki, K., et al. (2009). Updating quantum cryptography report ver. 1 (2009). arXiv:0905.4325. http://arxiv.org/abs/0905.4325

16. Dusek, M., Lutkenhaus, N., & Hendrych, M. (2006). Quantum cryptography. In *Progress in optics* (vol. 49, pp. 381–454). Amsterdam: Elsevier. https://doi.org/10.1016/S0079-6638(06)49005-3

17. Wolf, R. (2021). *Quantum key distribution. Lecture Notes in Physics* (vol. 988). Cham: Springer. ISBN 978-3-030-73990-4. https://doi.org/10.1007/978-3-030-73991-1

18. Niemiec, M., & Pach, A. R. (2012). The measure of security in quantum cryptography. In *2012 IEEE Global Communications Conference (GLOBECOM)* (pp. 967–972). https://doi.org/10.1109/GLOCOM.2012.6503238

19. Tang, X., Ma, L., Mink, A., Nakassis, A., Xu, H., Hershman, B., Bienfang, J., Su, D., Boisvert, R. F., Clark, C., & Williams, C. (2006). Quantum key distribution system operating at sifted-key rate over 4 Mbit/s. 6244:62440P. ISSN 0277786X. https://doi.org/10.1117/12.664455

20. Brassard, G., Salvail, L., Louis, S., Salvail, L., & Louis, S. (1994). Secret-key reconciliation by public discussion. In *Advances in cryptology - EUROCRYPT93* (vol. 765, pp. 410–423). Berlin: Springer. https://doi.org/10.1007/3-540-48285-7_35

21. Bennett, C. H., Bessette, F., Brassard, G., Salvail, L., & Smolin, J. (1992). Experimental quantum cryptography. *Journal of Cryptology, 5*, 3–28. ISSN 09332790. https://doi.org/10.1007/BF00191318

22. Ruth Ng Ii-Yung. A probabilistic analysis of Binary and Cascade. http://math.uchicago.edu/~may/REU2013/REUPapers/Ng.pdf, 2013. Online; Accessed July 7, 2022.

23. Sugimoto, T., & Yamazaki, K. (2000). A study on secret key reconciliation protocol. *IEICE Transactions on Fundamentals of Electronics, Communications and Computer Sciences, E83-A*(10), 1987–1991.

24. Mehic, M., Maurhart, O., Rass, S., Komosny, D., Rezac, F., & Voznak, M. (2017). Analysis of the public channel of quantum key distribution link. *IEEE Journal of Quantum Electronics, 53*(5), 1–8. ISSN 0018-9197. https://doi.org/10.1109/JQE.2017.2740426

25. Pedersen, T. B., & Toyran, M. (2015). High performance information reconciliation for QKD with CASCADE. *Quantum Information & Computation, 15*(5–6), 419–434.

26. Calver, T. (2011). *An empirical analysis of the cascade secret key reconciliation protocol for quantum key distribution*. Master Thesis.

27. Keath, C. (2010). *Improvement of reconciliation for quantum key distribution*. Ph.D. Thesis, Rochester Institute of Technology. http://arxiv.org/abs/1602.09140.

28. Yan, H., Ren, T., Peng, X., Lin, X., Jiang, W., Liu, T., & Guo, H. (2008). Information reconciliation protocol in quantum key distribution system. In *Proceedings - 4th International Conference on Natural Computation, ICNC 2008* (vol. 3, pp. 637–641). https://doi.org/10.1109/ICNC.2008.755

29. Martinez-Mateo, J., Pacher, C., Peev, M., Ciurana, A., & Martin, V. (2015). Demystifying the information reconciliation protocol cascade. *Quantum Information & Computation, 15*(5–6), 453–477. ISSN 15337146. https://doi.org/10.5555/2871401.2871407

30. Buttler, W. T., Lamoreaux, S. K., Torgerson, J. R., Nickel, G. H., Donahue, C. H., & Peterson, C. G. (2003). Fast, efficient error reconciliation for quantum cryptography. *Physical Review A, 67*(5), 052303. ISSN 1050-2947. https://doi.org/10.1103/PhysRevA.67.052303

31. Gallager, R. (1962). Low-density parity-check codes. *IEEE Transactions on Information Theory, 8*(1), 21–28. ISSN 0018-9448. https://doi.org/10.1109/TIT.1962.1057683

32. Elkouss, D., Martinez-Mateo, J., & Vicente, M. (2011). Information reconciliation for QKD. *Quantum Information & Computation, 11*, 226–238.

33. Elkouss, D., Martinez-Mateo, J., & Martin, V. (2013). Analysis of a rate-adaptive reconciliation protocol and the effect of leakage on the secret key rate. *Physical Review A - Atomic, Molecular, and Optical Physics, 87*(4), 1–7. ISSN 10502947. https://doi.org/10.1103/PhysRevA.87.042334

34. Elkouss, D., Leverrier, A., Alléaume, R., & Boutros, J. J. (2009). Efficient reconciliation protocol for discrete-variable quantum key distribution. In *2009 IEEE international symposium on information theory* (pp. 1879–1883). IEEE. https://doi.org/10.1109/ISIT.2009.5205475

35. Zhang, H.-F., Wang, J., Cui, K., Luo, C.-L., Lin, S.-Z., Zhou, L., Liang, H., Chen, T.-Y., Chen, K., & Pan, J.-W. (2012). A real-time QKD system based on FPGA. *Journal of Lightwave Technology, 30*(20), 3226–3234. ISSN 1558-2213. https://doi.org/10.1109/jlt.2012.2217394

36. Constantin, J., Houlmann, R., Preyss, N., Walenta, N., Zbinden, H., Junod, P., & Burg, A. (2017). An FPGA-based 4 mbps secret key distillation engine for quantum key distribution systems. *Journal of Signal Processing Systems, 86*, 1–15. https://doi.org/10.1007/s11265-015-1086-1

37. Yang, S.-S., Bai, Z.-L., Wang, X.-Y., & Li, Y.-M. (2017). FPGA-based implementation of size-adaptive privacy amplification in quantum key distribution. *IEEE Photonics Journal, 9*(6), 1–8. https://doi.org/10.1109/JPHOT.2017.2761807

38. Yang, S.-S., Lu, Z.-G., & Li, Y.-M. (2020). High-speed post-processing in continuous-variable quantum key distribution based on FPGA implementation. *Journal of Lightwave Technology, 38*(15), 3935–3941. https://doi.org/10.1109/JLT.2020.2985408

39. Elliott, C., Colvin, A., Pearson, D., Pikalo, O., Schlafer, J., & Yeh, H. (2005). Current status of the DARPA quantum network (Invited Paper). In E. J. Donkor, A. R. Pirich, & H. E Brandt (Eds.), *Proceedings of SPIE 5815, Quantum Information and Computation III* (vol. 5815, pp. 138–149). https://doi.org/10.1117/12.606489

40. Shimizu, K., Honjo, T., Fujiwara, M., Ito, T., Tamaki, K., Miki, S., Yamashita, T., Terai, H., Wang, Z., & Sasaki, M. (2014). Performance of long-distance quantum key distribution over 90-km optical links installed in a field environment of Tokyo metropolitan area. *Journal of Lightwave Technology, 32*(1), 141–151. ISSN 0733-8724. https://doi.org/10.1109/JLT.2013.2291391

41. Niemiec, M. (2019). Error correction in quantum cryptography based on artificial neural networks. *Quantum Information Processing, 18*(6), 174 (2019). ISSN 1570-0755. https://doi.org/10.1007/s11128-019-2296-4

42. Niemiec, M. (2011). *Design, construction and verification of a high-level security protocol allowing to apply the quantum cryptography in communication networks*. Ph.D. Thesis, AGH University of Science and Technology, Krakow.

43. Cachin, C. (1997). *Entropy measures and unconditional security in cryptography*. Ph.D. Thesis.

44. Wegman, M. N., & Carter, L. J. (1981). New hash functions and their use in authentication and set equality. *Journal of Computer and System Sciences, 22*(3), 265–279 (1981). ISSN 00220000. https://doi.org/10.1016/0022-0000(81)90033-7

45. Abidin, A. (2013). *Authentication in quantum key distribution: Security proof and universal hash functions*. Ph.D. Thesis, Linköping University.

46. Cederlöf, J., Larsson, J. A. (2008). Security aspects of the authentication used in quantum cryptography. *IEEE Transactions on Information Theory, 54*(4), 1735–1741. ISSN 00189448. https://doi.org/10.1109/TIT.2008.917697

47. Gilbert, G., & Hamrick, M. (2000). Practical quantum cryptography: A comprehensive analysis (Part one). arxiv:quant-ph, pp. 1–194.
48. Bennett, C. H. (1992). Quantum cryptography using any two nonorthogonal states. *Physical Review Letters, 68*(21), 3121–3124. ISSN 0031-9007. https://doi.org/10.1103/PhysRevLett.68.3121
49. Lomonaco, S. J. (1999). A quick glance at quantum cryptography. *Cryptologia, 23*(1), 1–41 (1999). http://arxiv.org/abs/quant-ph/9811056v1
50. Imre, S., & Balázs, F. (2994). *Quantum computing and communications.* Hoboken, NJ: Wiley. ISBN 9780470869048. https://doi.org/10.1002/9780470869048
51. Braunstein, S. L., & Pati, A. K. (2003). *Quantum information with continuous variables.* Dordrecht: Springer. ISBN 978-90-481-6255-0. https://doi.org/10.1007/978-94-015-1258-9
52. Li, J., Li, N., Zhang, Y., Wen, S., Du, W., Chen, W., & Ma, W. (2018). A survey on quantum cryptography. *Chinese Journal of Electronics, 27*(2), 223–228 (2018). ISSN 1022-4653. https://doi.org/10.1049/cje.2018.01.017
53. Qi, B., Zhu, W., Qian, L., & Lo, H.-K. (2010). Feasibility of quantum key distribution through a dense wavelength division multiplexing network. *New Journal of Physics, 12*(10), 103042.
54. Ralph, T. (1999). Continuous variable quantum cryptography. *Physical Review A, 61*(1), 010303. ISSN 1050-2947. https://doi.org/10.1103/PhysRevA.61.010303
55. Hillery, M. (2000). Quantum cryptography with squeezed states. *Physical Review A, 61*(2), 22309. https://doi.org/10.1103/PhysRevA.61.022309
56. Reid, M. D. (2000). Quantum cryptography with a predetermined key, using continuous-variable Einstein-Podolsky-Rosen correlations. *Physical Review A, 62*(6), 062308. ISSN 1050-2947. https://doi.org/10.1103/PhysRevA.62.062308
57. Einstein, A., Podolsky, B., & Rosen, N. (1935). Can quantum-mechanical description of physical reality be considered complete? *Physical Review, 47*(10), 777–780. ISSN 0031-899X. https://doi.org/10.1103/PhysRev.47.777
58. Cerf, N. J., Lévy, M., & Van Assche, G. (2001). Quantum distribution of Gaussian keys using squeezed states. *Physical Review A, 63*(5), 52311 (2001). https://doi.org/10.1103/PhysRevA.63.052311
59. Grosshans, F., & Grangier, P. (2002). Continuous variable quantum cryptography using coherent states. *Physical Review Letter, 88*(5), 57902. https://doi.org/10.1103/PhysRevLett.88.057902
60. Weedbrook, C., Lance, A. M., Bowen, W. P., Symul, T., Ralph, T. C., & Lam, P. K. (2004). Quantum cryptography without switching. *Physical Review Letter, 93*(17), 170504. https://doi.org/10.1103/PhysRevLett.93.170504
61. Grosshans, F., Van Assche, G., Wenger, J., Brouri, R., Cerf, N. J., & Grangier, P. (2003). Quantum key distribution using gaussian-modulated coherent states. *Nature, 421,* 238–241.
62. Lance, A. M., Symul, T., Sharma, V., Weedbrook, C., Ralph, T. C., & Lam, P. K. (2005). No-switching quantum key distribution using broadband modulated coherent light. *Physical Review Letter, 95*(18), 180503. https://doi.org/10.1103/PhysRevLett.95.180503
63. Pirandola, S., Mancini, S., Lloyd, S., & Braunstein, S. L. (2008). Continuous-variable quantum cryptography using two-way quantum communication. *Nature Physics, 4,* 726–730.
64. Garcia-Patron, R., & Cerf, N. J. (2006). Unconditional optimality of Gaussian attacks against continuous-variable quantum key distribution. *Physical Review Letter, 97*(19), 190503 (2006). https://doi.org/10.1103/PhysRevLett.97.190503
65. Leverrier, A., Grosshans, F., & Grangier, P. (2010). Finite-size analysis of a continuous-variable quantum key distribution. *Physical Review A, 81*(6), 62343. https://doi.org/10.1103/PhysRevA.81.062343.
66. Leverrier, A., Garcia-Patron, R., Renner, R., & Cerf, N. J. (2013). Security of continuous-variable quantum key distribution against general attacks. *Physical Review Letter, 110*(3), 30502. https://doi.org/10.1103/PhysRevLett.110.030502
67. Leverrier, A. (2015). Composable security proof for continuous-variable quantum key distribution with coherent states. *Physical Review Letter, 114*(7), 70501. https://doi.org/10.1103/PhysRevLett.114.070501

68. Bradler, K., & Weedbrook, C. (2018). Security proof of continuous-variable quantum key distribution using three coherent states. *Physical Review A, 97*(2), 22310. https://doi.org/10. 1103/PhysRevA.97.022310.
69. Jacobsen, C.S., Madsen, L. S., Usenko, V. C., Filip, R., & Andersen, U. L. (2018). Complete elimination of information leakage in continuous-variable quantum communication channels. *NPJ Quantum Information, 4*, 32.
70. Jouguet, P., Kunz-Jacques, S., & Diamanti, E. (2013). Preventing calibration attacks on the local oscillator in continuous-variable quantum key distribution. *Physical Review A, 87*(6), 62313. https://doi.org/10.1103/PhysRevA.87.062313
71. Huang, J.-Z., Kunz-Jacques, S., Jouguet, P., Weedbrook, C., Yin, Z.-Q., Wang, S., Chen, W., Guo, G.-C., & Han, Z.-F. (2014). Quantum hacking on quantum key distribution using homodyne detection. *Physical Review A, 89*(3), 32304. https://doi.org/10.1103/PhysRevA.89. 032304
72. Huang, J.-Z., Weedbrook, C., Yin, Z.-Q., Wang, S., Li, H.-W. Wei Chen, Guo, G.-C., & Han, Z.-F. (2013). Quantum hacking of a continuous-variable quantum-key-distribution system using a wavelength attack. *Physical Review A, 87*(6), 62329. https://doi.org/10.1103/PhysRevA.87. 062329
73. Qi, B., Lougovski, P., Pooser, R., Grice, W., & Bobrek, M. (2015). Generating the local oscillator "locally" in continuous-variable quantum key distribution based on coherent detection. *Phys. Rev. X, 5*(4), 41009. https://doi.org/10.1103/PhysRevX.5.041009
74. Lodewyck, J., Bloch, M., Garcia-Patron, R., Fossier, S., Karpov, E., Diamanti, E., Debuisschert, T., Cerf, N. J., Tualle-Brouri, R., McLaughlin, S. W., & Grangier, P. (2007). Quantum key distribution over 25 km with an all-fiber continuous-variable system. *Physical Review A, 76*(4), 042305. ISSN 1050-2947. https://doi.org/10.1103/PhysRevA.76.042305
75. Jouguet, P., Kunz-Jacques, S., Leverrier, A., Grangier, P., & Diamanti, E. (2013). Experimental demonstration of long-distance continuous-variable quantum key distribution. *Nature Photonics, 7*, 378–381.
76. Huang, D., Huang, P., Lin, D., & Zeng, G. (2016). Long-distance continuous-variable quantum key distribution by controlling excess noise. *Scientific Reports, 6*, 19201.
77. Leverrier, A., & Grangier, P. (2009). Unconditional security proof of long-distance continuous-variable quantum key distribution with discrete modulation. *Physical Review Letters, 102*(18), 180504. https://doi.org/10.1103/PhysRevLett.102.180504
78. Liao, Q., Guo, Y., Huang, D., Huang, P., & Zeng, G. (2018). Long-distance continuous-variable quantum key distribution using non-Gaussian state-discrimination detection. *New Journal of Physics, 20*(2), 23015.
79. ETSI ISG QKD. (2018). 007 - V1.1.1 - quantum key distribution (QKD); vocabulary. 1, 1–18.
80. Tomamichel, M., Lim, C. C. W., Gisin, N., & Renner, R. (2012). Tight finite-key analysis for quantum cryptography. *Nature Communications, 3*, 634. ISSN 2041-1723. https://doi.org/10. 1038/ncomms1631
81. Gisin, N., Ribordy, G., Tittel, W., & Zbinden, H. (2002). Quantum cryptography. *Reviews of Modern Physics, 74*(1), 145–195. ISSN 00346861. https://doi.org/10.1103/RevModPhys.74. 145
82. Langer, T. (2013). *The practical application of quantum key distribution*. Ph.D. Thesis, University of Lausanne.
83. Korzh, B., Lim, C. C. W., Houlmann, R., Gisin, N., Li, M. J., Nolan, D., Sanguinetti, B., Thew, R., & Zbinden, H. (2015). Provably secure and practical quantum key distribution over 307km of optical fibre. *Nature Photonics, 9*, 163–168. ISSN 17494893. https://doi.org/10. 1038/nphoton.2014.327
84. Mehic, M., Niemiec, M., Rass, S., Ma, J., Peev, M., Aguado, A., Martin, V., Schauer, S., Poppe, A., Pacher, C., & Voznak, M. Quantum key distribution. *ACM Computing Surveys, 53*(5), 1–41. ISSN 0360-0300. https://doi.org/10.1145/3402192

Chapter 2
Quality of Service Requirements

Modern telecommunications networks are based on packet-switching traffic processing and the methodology by which packets can be delivered using an arbitrary route. However, for some applications, merely finding the route to destination is insufficient. Applications can place multiple parameters in their requests, starting from the time it takes the network to deliver the packet and provide the response. Additionally, not all applications have the same priority, and it is necessary to clearly differentiate traffic. Quality of service (QoS) mechanisms are applied to enable networks to manage these requests.

In theory, QoS can be defined from multiple perspectives. In this book, QoS is approached from the end-user perspective of meeting expected levels of customer satisfaction during the use of network resources. Viewed from this aspect, when it is stated that the network provides good QoS, it means that the network is able to meet user requirements in a user-satisfactory manner. It also means that anything which impairs user satisfaction also falls under the definition of QoS. Here, we specifically refer to reliability, robustness, security and other aspects which, because of their simplicity, are most often viewed independently of QoS.

It is often stated that QoS consists of two components: the QoS of the network and the end-user application. The end-user QoS component is determined from the software application aspect through utilization of network resources. Since application implementations do not fall directly under networking, they are not discussed in detail in this book. This book discusses the QoS network component: how does a network meet defined service requirements? In this chapter we describe common approaches for implementing QoS in IP networks and methods of combining these mechanisms in QKD networks.

© Springer Nature Switzerland AG 2022
M. Mehic et al., *Quantum Key Distribution Networks*,
https://doi.org/10.1007/978-3-031-06608-5_2

2.1 Quality of Service

Before the introduction of Internet Protocol (IP), networks had limited capacity and applications had to adapt to assigned network resources with limited configuration options. The parameters by which network services differed were significantly limited, which in turn limited the development of new telecommunications services. Privatization and opening of telecommunications markets commenced in the 1980s. Internet Service Providers (ISP) looked for new ways of attracting customers and providing better network conditions. This gave rise to the differentiation of network traffic according to priority. In this context, the network increasingly attempts to adapt to the requests of network applications. New QoS approaches were developed during this period, some which are still in use today.

According to the ITU-T recommendation, Quality of Service (QoS) refers to the "totality of characteristics of a telecommunications service that bear on its ability to satisfy stated and implied needs of the user of the service" [1]. The QoS concept strictly relates to satisfying the primary parameters of specific applications or traffic during an active connection between a generic source and a destination.

We seek to separate the definitions of QoS, QoS constraints and QoS mechanisms. QoS is an aim; QoS constraints are precise measurable values which describe the aim, while QoS mechanisms are the means applied to achieve that aim. Some views allow the QoS approach to be neglected by increasing the network capacity. If the network is overprovisioned, sufficient resources are available for all requests. Additionally, the network is thus prepared for future upgrades and services. Although at first view such an approach seems an elegant solution, increasing network capacity is not an option in a dynamic network such as a QKD network.

2.2 Quality of Service Constraints

By default, the network will treat all users equally and provide them with the *BestEffort* level of service. The network will do everything it can to deliver the packet from the source to destination as quickly as possible but without any given guarantees. With the development of new network services, time-sensitive multimedia applications such as voice over IP (VoIP), video conferencing, and others are emerging. These applications require the network to perform service differentiation and consider additional parameters other than the mere delivery of packets from source to destination. It is therefore necessary to define the QoS parameters and required values whose fulfillment is expected from the network.

The required constraints can be specified either according to user's preferences (e.g., bounds on delay while managing interactive multimedia communications, preferred data-rate during download operations, etc.) or commercial requirements (e.g., the need to complete a transaction within given time constraints, the minimum ensured amount of resources during wideband transmission, etc.). QoS requirements or constraints can be described in either "relative" or "absolute" terms. In the first case, the requirements refer to the treatment received by a specified class of packets. By contrast, in the second case, the constraints provide some exact metrics (e.g., delay, loss, jitter, etc.) as upper/lower bounds or statistical percentages. For the sake of argument, an example of a statistical bound could be "during transmission, the maximum packet loss will be no more than 3%" or an upper bound could be "no packets will be subject to an end-to-end (e2e) delay greater than 200 ms".

   Different applications have different requirements, and in practice, the user should be able to specify the QoS level required. For example, for telephone or videophone services, the end-to-end delay is crucial. An upper limit for a one-way delay according to the ITU-T's G.114 recommendation guidelines is 150 ms. This end-to-end delay includes the total time taken to capture, digitize and encode/compress the audio/video data, encrypt and transport it from the source to the destination, decrypt, decode and display it to the user. The default G.729 codec requires packet loss to be significantly less than 1% to avoid audible errors [2]. Additionally, VoIP calls require constant bandwidth, which depends on the codec applied. Some parameters are therefore more important for communication than others, which means some parameters can tolerate fluctuations without affecting communication quality. Data transfers (FTP/HTTP downloads) may tolerate massive delays and higher packet-losses (given that retransmissions recover them), as illustrated in Table 2.1. In these terms, when an ISP has to manage its customers, they can agree about a contract regarding a given set of terms, often called the Service Level Agreement (SLA).

   Sometimes, it is more intuitive and convenient for applications to specify QoS requirements and the required guarantee level. Generally, three QoS levels concerning guarantees are distinguished [3]:

- Hard or deterministic, which implies that the requirements must be met in full,
- Soft or statistical, where specified QoS requirements should be met to a certain specified level,
- Best effort, where no guarantee is provided at all.

Table 2.1 Typical QoS requirements for different types of application

Traffic type	Sensitivities			
	Bandwidth	Loss	Delay	Jitter
Voice	Very low	Medium	High	High
E-commerce	Low	High	High	Low
Transactions	Low	High	High	Low
E-mail	Low	High	Low	Low
Telnet	Low	High	Medium	Low
Casual browsing	Low	Medium	Medium	Low
High Quality browsing	Medium	High	High	Low
File transfer	High	Medium	Low	Low
Video conferencing	High	Medium	High	High
Multicasting	High	High	High	High

The application can thus determine which QoS parameter should be strictly met and whether variations in the values of another QoS parameter can be tolerated and to what extent. In the context of QKD networks, this view is fundamental due to the overlapping understanding of some of the QoS parameters discussed in the remainder of this book.

QoS parameters can be roughly classified into additive (e.g., delay) and min-max (e.g., available bandwidth). In the case of an additive parameter, the QoS value of a path is equal to the sum of the corresponding weights of the links along that path. For min-max parameters, the QoS value of a path is the minimum (or maximum) link weight along that path. It should be noted that there are also multiplicative parameters in which the QoS parameter along the path is equal to the product of QoS parameters of the links (e.g., loss probability). However, by taking the logarithm of multiplicative parameters, the problem reduces to additive parameters.

QoS guarantees can be referred to either over a group of packets transmitted into a given period of time (flow) or to an aggregate of flows (aggregate QoS). Although the introduction of packet-switching network organization allows packets to travel different paths, network management mechanisms most often observe network traffic through flow mapping. The aim is to map traffic flow using clear determinants such as source and destination addresses, types of application, ports on the transport layer, etc. Accordingly, different network flows within a network with different processing prioritizations can be distinguished:

- Elastic traffic, typically referred to as data traffic, in which the transmitted information is not time-sensitive but requires eventual proper delivery such as email, web browsing, file transfer or similar;
- Inelastic traffic, typically referred to as real-time traffic, for which the transmitted information is only useful if it is received within a given time constraint, for example, VoIP, IPTV, video conferences, time-dependent interactive applications or similar.

The most important aspect of inelastic traffic is that a message generated by the source device in an application is time-sensitive. The destination device must receive it within a specified time window. Let us suppose that the time between the generation of the message at the source device and its reception at the destination device, which is defined as delay, exceeds this time constraint. In this case, the message is considered lost, regardless of whether it is ever received at the destination. Also, it is possible that the aggregate rate of the input traffic to the network (or a portion of the network) temporarily exceeds the network's capacity, in which case packets may experience extended delays or be dropped by the network nodes. This is referred to as network congestion, and the network needs to adopt corrective QoS mechanisms to return the network load to an optimal state so that other users are not harmed. Additionally, QoS mechanisms can be used proactively to avoid congestion [4].

To provide QoS, a network needs to perform:

- Differentiation between classes of traffic to clearly distinguish serving priorities,
- After distinguishing the traffic classes, the network needs to be able to treat these classes distinctly by providing resource assurance and service differentiation.

2.3 Quality of Service Components

Different applications react differently to the starvation of resources. A real-time multimedia stream can ignore late received packets, while content download applications can tolerate delays. To decide whether sufficient resources are available, the current network load needs to be known. For predicted services, the estimated network load is based on measurement; for guaranteed services, it is based on the pre-specified characterization of connections which are active in the network.

The network needs to implement some of the logical QoS components which will provide the functionality to meet the stated requirements. The combination of components allows the definition of different QoS architectures (some of them are discussed in the following sections). The most common QoS entities are:

- **Traffic classification**: This has the task of separating network traffic based on predefined determinants such as the source or destination address, port number, protocol type or other type of identifier. The aim is to identify traffic which needs higher priority service and richer treatment.
- **Traffic marking**: This is a logical continuation of the classification with the aim of clearly marking the filtered traffic. An analogy often applied is road traffic and rotating lights of different color used to mark high-priority vehicles (police, firefighters and others). In telecommunications networks, marking can be accomplished with various methods, from rising flag bits in packet headers to adding completely new headers. Marked packets are detected quickly in the rest of the network and additionally speed up the reclassification process.
- **Metering**: Network nodes can implement dedicated tools to detect and monitor the characteristics of network flows. The obtained information can be used in the traffic marking process or the decision-making process while processing network traffic. Some examples of traffic measurement tools are token and leaky bucket.
- **Policing**: If the amount of traffic in the network exceeds the defined limits, the network can expand metering tools with defensive actions in the context of policing (packet dropping). Policing measures should not only be adopted in the context of excessive network traffic, they can also be used to prevent the occurrence of traffic side effects such as burstiness.
- **Shaping**: Similar to policing techniques, shaping techniques aim to shape network traffic into the desired shape. They are most often accomplished in a manner that the network traffic is intentionally delayed to be shaped according to the desired mold at the output interface.

- **Queuing**: This method stores packets received for processing in dedicated queues. Different techniques for selecting marked packets from queues which take into account traffic priority, network capacity and fairness aspects of service can be selected.
- **Scheduling**: This includes techniques for controlling the capacity of queues. Specifically, the techniques are used to define which packet should be removed from which queue. In performing these techniques, the mechanisms may also consider traffic priorities and fairness of service.
- **Switching**: Selected packets which meet the requirements of previously listed modules can be served. An appropriate route which directs them to the destination is determined.
- **Admission control**: Depending on the QoS architecture, networks may implement dedicated entities to explicitly decide whether cross-border traffic can be accepted. These entities seek to prevent further processing steps in the early stages based on consideration of the requested requirements and current network utilization.

The QoS solution designer will most often resort to a combination of some of the listed QoS blocks to form a QoS architecture (or often called a QoS model). A generic router structure with included QoS components is shown in Fig. 2.1. In Chap. 3, we consider several QoS models and the possibilities of their integration with QKD networks.

A vision to meet the QoS requirements significantly affects other network entities in which requests for service prioritization are also considered:

- **QoS MAC**: Represents the processes of collecting up-to-date network status information so that a decision on the possibility of fulfilling QoS requirements can be made. The Media Access Layer (MAC) layer is the most frequently addressed because it represents the connection between hardware components and higher logic layers.

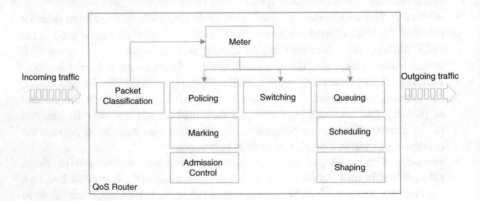

Fig. 2.1 A generic router with QoS components

- **QoS signaling**: Aims to implement provisioning methods, or more precisely, to consider whether resources are available for implementation of the service with the required QoS parameters by analyzing different network paths. Signaling techniques can also be used to reserve resources for selected resources, renew and ultimately schedule or tear down reservations.
- **QoS routing**: Aims to find the optimal path for delivering packets to the destination (if any). When defining a path decision, unlike standard routing protocols, QoS routing can consider multiple peer parameters. Also, when making a routing decision, it may be influenced by already established reservations or information obtained from lower network layers.

2.4 QKD Networking

In contrast to IP networks where the link state information such as delay, bandwidth, cost, and loss rate are implied, QKD networks require additional information to ensure that QoS requirements are met. In this section, we consider the basic features of a QKD system to denote those QoS requirements. The following chapters discuss methods of organizing QoS models and mechanisms which satisfy QoS constraints in QKD networks.

A QKD link, or simply "link", denotes a logical connection between two remote QKD nodes. The link consists of a quantum channel used to exchange photons and a public channel for post-processing operations (Fig. 2.2). The existence of these two channels limits the key generation rate and distance range of the link. The link can be established between parties connected by a direct optical fibre or a free line of sight. However, it is important to keep in mind that the purpose of a QKD link is to provide an ITS level of security. Therefore, the process of establishing cryptographic keys is based on the laws of quantum physics instead of the mathematical complexity of an algorithm [5, 6].

In practice, multiple quantum or public channels can be used to implement a single logical QKD link with improved performance. Additionally, the quantum channel requires the implementation of a (time-stable) synchronization channel [7]. It is used to deliver time references and can be physically separated from the optical

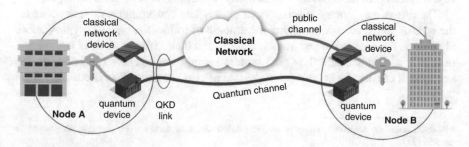

Fig. 2.2 The logical QKD link consists of a public (blue line) and a quantum channel (red line)

fibre through which the quantum channel is accomplished. However, observed from higher networking layers, the synchronization channel can be considered a part of the process which takes place through the quantum channel, and therefore it is often not considered in the performance analysis of QKD links.

A QKD link's performance is higher when it is accomplished through a dedicated optical channel often referred to as a dark fibre. Nevertheless, the high rental or deployment costs of optical fibres can limit their implementation and the amount of time used in practice. Instead of fibre, a free space link with lower performance can be used. Although these allow the distance limit to be overcome, QKD free space systems provide a lower key rate because they depend on parameters such as the visible light path and suitable atmospheric conditions. The free space connection is indispensable in establishing satellite QKD systems, for example, the Chinese quantum Micius satellite, which has successfully demonstrated QKD connection at distances of 645 to 1200 kilometers [8]. The same category includes approaches for the implementation of QKD systems with temporary overflight, such as links with flying aircraft [9, 10], balloons [11], drones [12] and other methods [10, 13–15].

It is important to mention the approaches for integrating quantum channels into existing communications network architecture. Installation costs can be reduced by combining the quantum and public channels over the same fibre using wavelength division multiplexing (WDM) techniques [16, 17]. Recent results have shown that CV-QKD systems have the potential to be integrated over WDM channels because they are more tolerant to resulting noise [18–20].

Regardless of the type of QKD system, there is a limitation that the link length decreases significantly with increasing path losses or optical detector noise. Although the dark count[1] values can be considered constant for the given detector and its settings, the key rate of the QKD system directly depends on the distance of the links. Commercial QKD solutions are generally limited to a few hundreds of kbps, while link distance is roughly limited to 100–150 km. Although results for longer distances have been reported, they have generally demonstrated lower generated key rates and vice versa [22–24].

Because of the limitations which reduce the ability to generate large amounts of keys as needed, QKD systems tend to implement key storage on both sides of the link. The aim is to produce keys in advance and store them for future use. As shown in Fig. 2.3, the link is usable only when a key is available for use, i.e., a QKD link without available keys can be viewed as "currently unavailable", because without a key, it cannot provide its basic functionality, which is ITS communication [25].

The rate of key storage discharge, or simply key consumption rate, depends on the type of application determining the amount of traffic, the type of cryptographic algorithms for encryption and authentication, and their settings, such as key refresh rate. If it is assumed that the QKD system settings are constant (distance, device settings), it can also be assumed that the rate of new key generation is constant. This

---

[1] A dark count is an event where a single-photon detector clicks even though no photon is present [21].

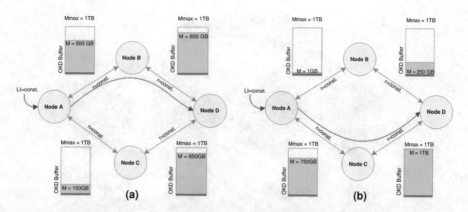

Fig. 2.3 A simple example of routing with consideration of the amount of key material in key storage. In the case of communication between nodes A and D, node A needs to make a decision on choosing the best path A-B-D or A-C-D, which depends on the state of network links. Initially node A chooses the path A-B-D (left) and switches to path A-C-D (right) when the path A-B-D becomes unavailable because of a lack of key material

means that key buffers will be filled at a steady rate, often referred to as the key charging rate, which is an important determinant in defining the QoS mechanisms discussed below.

2.4.1   QKD Networks

To accomplish ITS secure communication over a greater distance and support communication with multiple users, various approaches in QKD network organization have been considered. All approaches commonly view QKD network nodes as secure entities in the context of unlimited processing power or power supply. It is therefore assumed that the nodes are secure network entities and that a security threat to one of the network nodes can jeopardize the security of the entire network. The concepts of multipath communication and network coding can be used to avoid this assumption [26], as discussed in Chap. 7.

Different approaches to accomplish QKD communication have been considered since the first demonstration of a practical QKD network. The DARPA QKD network which was fully operational in early 2003 established the principle of key relay architecture [27]. The key relay approach is often referred to as BBN key relay, because in the current world of QKD networks the term key relay possesses a different meaning.

Figure 2.4 depicts the BBN key relay principle for secure transfer of message n between nodes A and D. Let us suppose that node A is a node with the highest network identifier ID. This node will calculate the route to the distant node D and initiate the reservation requests to all nodes on the path (Fig. 2.4a). The authors did

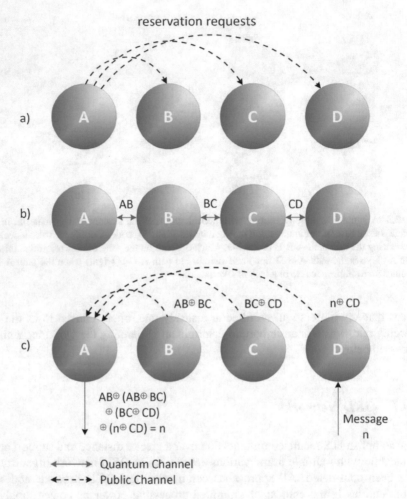

Fig. 2.4 The BBN key relay principle

not specify ways of maintaining the reservation. If an intermediate node is not able to supply enough keys, it will propagate the failure error message to all other nodes along the path releasing previously reserved keys. If the key reservation request is successful, each node negotiates with its neighbors the key that will be used for secure transfer of the message (Fig. 2.4b). As shown in Fig. 2.4c, in the last step each node will respond to node A with XOR values of the generated keys to its first neighbors. The last node D will send the XOR of the desired message n and the key which was established with the first neighbor in the defined path. Finally, node A will perform XOR of all received values and extract the message n.

The message exchange described in steps (a) and (c) takes place through a public channel. Therefore, the algorithm needs to implement additional tools to verify whether the messages sent were indeed from nodes in the path and not from an attacker.

During 2004, the European Commission (EC) funded the FP6 project Secure Communication based on Quantum Cryptography (SECOQC) (Secure Communication based on Quantum Cryptography). The project included 41 research and industrial partners from 11 EU countries, Russia and Switzerland. SECOQC sought to demonstrate applications of QKD technology and to define the foundations of future QKD communication by analyzing the methods of practical implementation of QKD systems. Approaches to integrating QKD technology into existing IP networks were also considered, and a hop-by-hop approach was proposed [28].

Unlike the BBN key relay principle, which explicitly involves queries to reserve keys and process all received messages at the destination hub, the hop-by-hop approach is based on simply forwarding encrypted messages along the path. As shown in Fig. 2.5 the hop-by-hop approach involves encrypting the message at the source node, directing it to intermediate nodes where the authentication tags are first verified and the message is decrypted, and re-encrypting with a key for the next node [28].

Hop-by-hop connection can be considered the serial connection of separate QKD systems, the main drawback being the time needed to transfer data along the path (congestion at one node would slow down overall communication). If the BBN key relay approach is viewed in terms of performance, it can be seen as the parallel transmission of partial components of the overall message. BBN key relay requires both high computational power and communication resources at node A. If the number of messages to be transmitted increases significantly as does the number of nodes in the path, communication can grow into a DoS attack and present practical challenges in processing the received data. However, network performance might be faster because parallel data transmission is used. Regardless, an additional weakness exists: if one of the nodes is affected by congestion, the entire decryption process will also be impeded.

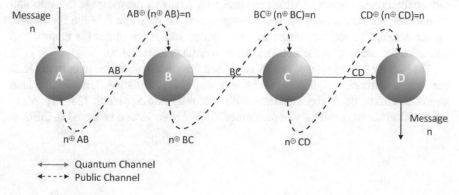

Fig. 2.5 The hop-by-hop principle

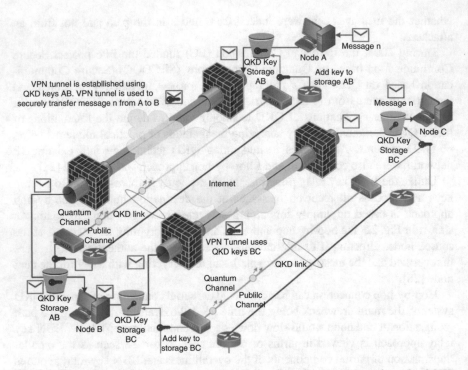

Fig. 2.6 Illustration of a hop-by-hop connection. Yellow lines denote quantum channels of QKD links implemented in a direct point-to-point connection over a limited distance. Message *n* is encrypted and decrypted in hop-by-hop manner on all nodes along the route A-B-C

As shown in Fig. 2.6, the hop-by-hop approach assumes the establishment of $k - 1$ VPN tunnels between k nodes for secure message transfer. Additionally, it is necessary to maintain these tunnels as long as the message transfer is active.

In addition to the transmission of encrypted messages, one of the objectives is to reduce the number of nodes which are generally aware of the message transmission path. If the attacker knows the route along which the message is being forwarded, he can attempt to redirect traffic to those nodes which are under his control and endanger the security of the network (more details in Sect. 2.4.5).

The hop-by-hop approach and its variations are widely accepted in current QKD network organization. QKD networks in Europe, Japan, China, UK, Switzerland and others have been accomplished with a hop-by-hop approach [28–31]. More information about previously implemented QKD networks can be found in [32].

2.4.1.1   Key Relay

From the previously described approaches, we can observe that the availability of all links is required for the transmission of message n. Often, this may not be practical due to the following:

- All QKD links and associated VPN tunnels must be available at all times, requiring significant financial allocation to reserve communications resources (optical fibres, QKD devices, and others).
- The availability of keys along the route might delay the message transfer process. In the worst case, the transmission may be canceled if it is not possible to find a route with enough keys to transmit message n to the destination node.

To overcome these issues, we can implement a key relay concept based on the transfer of key k instead of message n, as shown in Fig. 2.7. The origin of key k can be from a Quantum Random Number Generator (QRNG) on the source node. Message transfer is secured using key k through the public network infrastructure, meaning that message transfer security is based on the secrecy of key k, while the QKD network provides a secure infrastructure for the transfer of key k. It also means that rate of QRNG may impact the end-to-end key rate.

Another approach is to take advantage of the QKD features since the QKD process is a known key establishment process. The key can be based on the randomness of quantum components, and the initial key AB can be relayed to the destination node. This approach does not require the use of QRNG, reducing the cost of implementation. As shown in Fig. 2.8, the final key rate of the end-to-end path A-D is limited by the key rate of link A-B since the reduction of key rates on other links after node B can be avoided using alternative routes.

Using the key relaying approach, communication resources can be leased for a limited time and keys can be generated in advance, i.e., the node A does not have to maintain a VPN connection with node B. An additional buffer will be implemented to store the keys relayed between nodes A and C. Later, the relayed keys can be

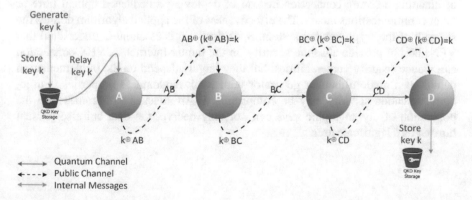

Fig. 2.7 Relaying of random key k

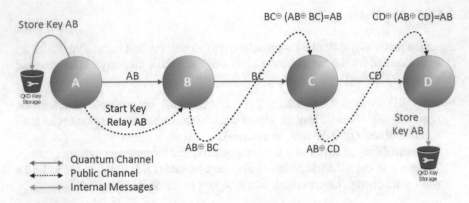

Fig. 2.8 Relaying of the QKD key

used as needed. As shown in Fig. 2.9, if enough keys have been relayed, a direct VPN tunnel can be established between A and C to securely transfer the message n.

2.4.2 QKD Virtual Private Networking

The convergence of QKD technology depends on its acceptance in today's widespread IP networks. Virtual Private Network (VPN) techniques for secure communication in IP networks have therefore been expanded to support QKD-generated keys. Unlike the previous sections which consider the establishment and distribution of keys, this section discusses the use of the key, i.e., key consumption for encryption purposes.

VPN is a virtualization technique implemented over a physical infrastructure to ensure secure data transfer between remote entities. It can be deployed over an unsecured public network such as the Internet, providing an elegant means of attaining a secure connection instead of deploying a dedicated optical fibre or other communications media. The effectiveness of the applied algorithm defines the strength of the cryptographic system. Specifically, it is essential to understand that VPNs do not provide absolute security on the public Internet. A VPN connection can reduce security vulnerabilities, but they largely depend on the implementation details. If a random number generator with weak randomness properties is used, communication security may be compromised. An attacker who comes into the possession of cryptographic keys can not only decrypt the data but also present himself as a legitimate user.

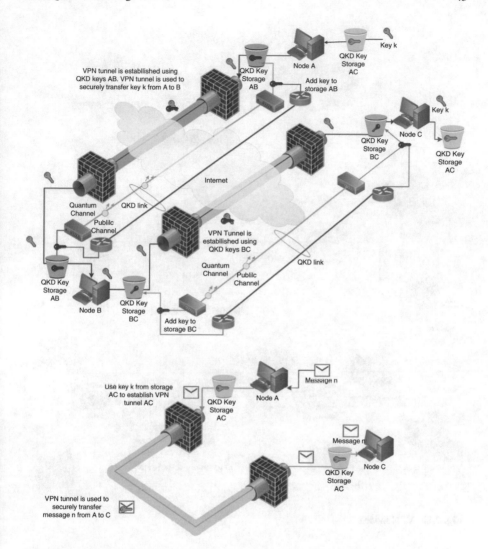

Fig. 2.9 Illustration of key-relay connections. Yellow lines denote quantum channels of QKD links implemented in a direct point-to-point connection over a limited distance. Key k is relayed in hop-by-hop manner along the route A-B-C. Message n is protected using the previously relayed key k

It is therefore necessary to understand the approaches for integrating QKD keys into VPN connections to reduce the risk of security threats. The primary purpose of QKD technology is ITS communication, and suitable application of QKD keys can significantly reduce these risks. As shown in Fig. 2.10, a VPN connection can be modeled in three different ways:

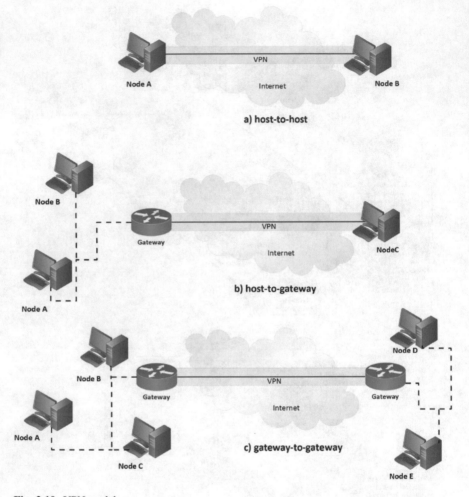

Fig. 2.10 VPN models

- **host-to-host** model is the least used model in VPN connections. It provides a direct connection between two remote network entities. Although this model establishes secure end-to-end communication, it demonstrates shortcomings in the scalability and management of a large number of VPN connections.
- **host-to-gateway** model allows the integration of a single remote network node with a remote network over a public network infrastructure. A VPN gateway can be a dedicated network device, and remote network users can log in as needed. It is also important to note that end-to-end communication is not secured, as a VPN does not provide additional security mechanisms within the local network.
- **gateway-to-gateway** model is used to connect a group of remote network nodes over public infrastructure. The two end-point VPN entities (gateways) of each remote network establish a secure VPN connection. The overall communication

between the network entities from these networks is accomplished through a secure VPN tunnel. Nevertheless, this model is the easiest to implement since the established VPN tunnel is transparent to network nodes within remote networks. No new VPN connection needs to be implemented since one already exists between the gateways. Still, it must be emphasized that communication within the local network is not secure.

2.4.3   IPsec

Although multiple techniques can be applied to attain a VPN connection, in this section, we are particularly interested in the Internet Protocol security (IPsec) framework, which is predominantly used in QKD networks. IPsec is a suite of network security control solutions which ensure private communication over IP based networks. Depending on the implementation and configuration method, IPsec can provide:

- **Confidentiality**: IPsec encrypts data and unauthorized parties are unable to read the exchanged messages.
- **Integrity**: IPsec generates message authentication checksum values to ensure no data has been altered during the transit.
- **Replay Protection**: IPsec uses sequential sequence numbers for each packet and a sliding window of permitted sequence numbers. The replayed packet which has sequence numbers outside the permitted window will be discarded, which protects against DoS attacks.
- **Access Control**: IPsec supports integration with authentication servers such as a RADIUS or TACACS. Those can be used to filter authorized IPsec users that have permission to access specific network resources.

The following sections provide a high-level and schematic overview of IPsec components, while a detailed depiction can be found in [33]. These sections serve to refresh the knowledge of IPsec fundamentals and to better understand the methods of combining IPsec with QKD, which is discussed in Sect. 2.4.4.

2.4.3.1   Authentication Header

One of the basic IPsec components is the authentication header (AH) used to provide integrity and data authentication. As defined in RFC 2402, AH works in tunnel or transport mode (Fig. 2.11) [34]. Tunnel mode is used in VPN connections that involve a gateway entity. As shown in Fig. 2.11b, the original IP header is encapsulated using a new IP header. The gateway of the corresponding VPN tunnel is in charge of decapsulating the packet.

As defined by RFC 3884, transport mode is mainly used in a host-to-host VPN connection without modifying the IP header or creating the new one [35]. Integrity is

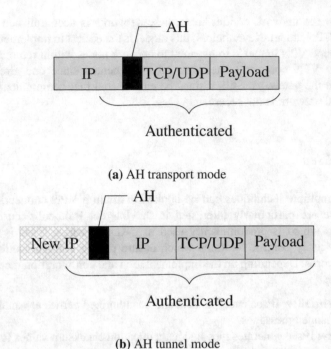

(a) AH transport mode

(b) AH tunnel mode

Fig. 2.11 IPsec AH protocol packet header organization. (**a**) AH transport mode. (**b**) AH tunnel mode

based on calculation of the hash value using a keyed Hash Message Authentication Code (HMAC) algorithm.[2] Since some IP header values are dynamic, such as the time to live (TTL) or IP header checksum, those fields are ignored when the hash is calculated. However, AH protocol includes the source and destination IP address in the HMAC input, which can cause issues when traffic is processed by Network Address Translation (NAT) devices. Since NAT might change these fields (i.e., changing the source IP address from private to public), the calculated AH integrity hash values will not match.

An AH header consists of six fields, which are depicted in Fig. 2.12:

- **Next Header**: the field is used to carry information about the next header in the header chain. In tunnel mode, the next header is the IP packet, hence the field's value is 4. In transport mode, the field carries the next transport-layer protocol's value, such as TCP (value 6) or UDP (value 17).
- **Payload Length**: the field contains the length of the payload.
- **Reserved**: the field is reserved for further use. The default value is 0.

---

[2] The difference between a standard and keyed hash algorithm is that the former generates a hash value based only on a message. The latter generates the hash value using both a message and secret key.

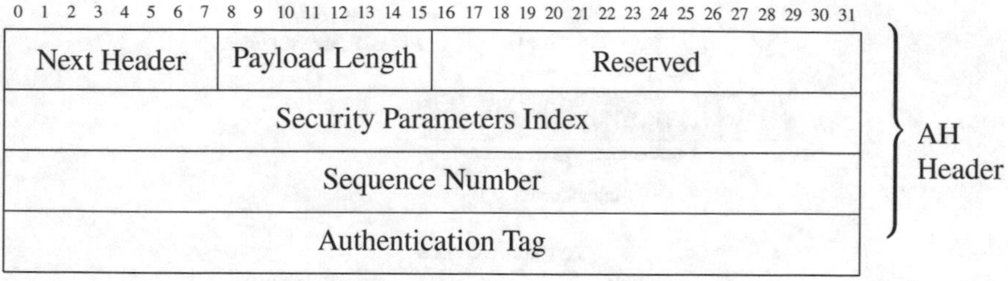

0 1 2 3 4 5 6 7 8 9 10 11 12 13 14 15 16 17 18 19 20 21 22 23 24 25 26 27 28 29 30 31

Next Header	Payload Length	Reserved
Security Parameters Index		
Sequence Number		
Authentication Tag		

AH Header

Fig. 2.12 Structure of an AH header

- **Security Parameter Index**: the randomly generated 32-bit identifier used on the receiving side of the IPsec tunnel. The SPI value is unique identifier which allows the used Security Association (SA) to be determined if multiple SAs are established between peers.
- **Sequence Number**: the sequential identifier of the packet. It is used to prevent replay attacks because duplicate packets have the same sequence number.
- **Authentication Tag**: the hash value calculated using the HMAC algorithm.

The AH protocol is rarely used since it does not provide confidentiality due to the lack of an encryption algorithm. To enable confidentiality, IPsec implements the ESP protocol.

2.4.3.2 Encapsulating Security Payload

The Encapsulating Security Payload (ESP) can be used to provide data integrity, data origin authentication, anti-replay protection, and optionally, encryption protection. It also works in transport and tunnel modes which operate in a similar manner to AH [36].

As shown in Fig. 2.13a, the original IP header is used without modification in ESP transport mode. This can provide incompatibility with NAT, and therefore, ESP transport mode is mostly used in host-to-host architecture. The TCP checksum is formed over TCP and IP fields. But, because of the inability to decrypt the encrypted TCP header, NAT cannot recalculate the checksum. The ESP tunnel is the most popular IPsec mode. A new IP header with the IP addresses of the ESP endpoints (gateways) is prepended to the packet (Fig. 2.13b). The original IP header and the payload are encrypted. In the ESP tunnel mode, NAT is not aware of the encrypted TCP header since the first next header in the header chain is the ESP header. Therefore, NAT will not attempt to recalculate the checksum.

The ESP includes an ESP header and an ESP trailer. The header consists of two fields:

- **Security Parameter Index (SPI)**: the randomly generated 32-bit identifier used on the receiving side of the IPsec tunnel. The SPI value is unique identifier which

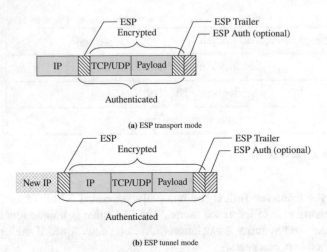

Fig. 2.13 IPsec ESP protocol packet header organization. (**a**) ESP transport mode. (**b**) ESP tunnel mode

allows the used Security Association (SA) to be determined if multiple SAs are established between peers.

- **Sequence Number**: the sequential identifier of the packet. It is used to prevent replay attacks because duplicate packets have the same sequence number.

The ESP protocol uses symmetric cryptography for the encryption of IP packets. The payload of the packet is encrypted whereas the Initialization Vector (IV) is exchanged unencrypted. The purpose of a random IV value in the encryption process is to provide a different encrypted value even if the packet's payload is identical. As long as the IV is random and not under the attacker's control, the value can be exchanged unencrypted.

The ESP trailer consists of the following fields (Fig. 2.14):

- **Padding**: additional bytes to fill the payload's missing parity space. Because ESP uses block ciphers, padding bytes are needed to ensure that the encrypted data corresponds to the block size used for encryption.
- **Authentication Tag Field**: the hashed authentication value.
- **Padding Length**: the indicator of included padding bytes (if any).
- **Next Header**: the field is used to carry information about the next header in the header chain. In tunnel mode, the next header is the IP packet, hence the field's value is 4. In transport mode, the field carries the next transport-layer protocol's value, such as TCP (value 6) or UDP (value 17).

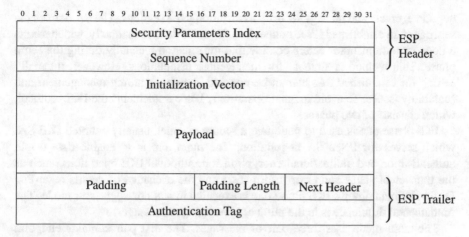

Fig. 2.14 Structure of ESP header and ESP trailer

2.4.3.3 IP Payload Compression Protocol (IPComp)

Adding an AH/ESP header may have an impact on communications performance due to increased overhead [37]. IPsec introduces lossless compression by implementing the IP payload compression protocol (IPComp), as defined in RFC 3173 [38]. Since the task of any encryption algorithm is to make the data appear random to the observer, compression algorithms are used before encryption. IPComp focuses on payload, and its efficiency depends on the data being transferred, i.e., for packages with a small payload, it will provide low compression efficiency. Additionally, some data may already be compressed on the application layer and hence the use of additional compression may be redundant since it introduces additional computational costs.

Because of the above, IPComp will analyze the effect of compression before generating the output value. If no benefit is achieved in applying compression, the original non-compressed packet will be sent so that at least the receiving party conserves resources by not performing decompression.

2.4.3.4 Internet Key Exchange (IKE) Protocols

After a brief description of VPN communication, we discuss how IPsec can be combined with QKD networks. Here we provide an overview of the IKE protocol, while more details of the numerous variants and extensions of IKE protocol can be found in [33, 39].

The IKE protocol was initially proposed in RFC 2409 as an extension of Internet Security Association and Key Management Protocol (ISAKMP) [40]. In IKE, the ISAKMP framework for authentication and independent key exchange is combined with SKEME [41] and Oakley (RFC 2412 [42]) key agreement protocols to negotiate, create and manage security association (SA) sessions. The SA defines the settings of an established IPsec connection. It can be defined manually, which means a network administrator enters confidential information manually during the setup phase. This method is suitable for manageable installations. However, manually setting the established SAs cannot be updated, thereby entailing management and scalability issues. To avoid manual operations, IPsec implements the IKE protocol, which consists of two phases.

IKE phase one is used to negotiate a secure channel, usually denoted IKE SA, which serves for IPsec SA negotiation. The main aim is to enable data origin authentication and bidirectional encryption for additional IKE operations, such as the transfer of status and error information and the exchange of details regarding Diffie-Hellman groups. IKE phase one is executed in *main* or *aggressive* mode. The fundamental difference is in the number of exchanged messages.

The main mode uses three pair of messages. The first pair contains endpoint proposals of common parameter combinations which can be used to form the SA, for example: encryption algorithm (i.e., AES-CBC), keyed hashing algorithm (i.e., HMAC-SHA), authentication method, and Diffie-Hellman key agreement primitive (DH) group (i.e., 2048-bit MODP DH group). In the context of this book, the DH group parameter determining the key length and encryption generator type (EC2N, MOPD, $G[2^N]$) is our main interest. The first pair of messages also include the exchange of cookies, similarly to those of the Oakley protocol. The cookies are used to record the endpoint's IP address and a time-limited counter to protect against DoS attacks. The second pair of messages use the agreed parameter from the previous step to perform a DH key exchange and form a shared secret, usually denoted Session Key ID (SKEYID). This value depends on the authentication method, which in the IKE process has a crucial role, i.e., in the first pair of IKE messages (messages [01] and [02] in Fig. 2.15) an authentication method is agreed, which may be based on:

- **Pre-shared keys**: the use of pre-shared symmetrical keys. In this case

$$SKEYID = prf(pre\_shared\_key, N_i|N_r),$$

  in which prf is a pseudorandom function [43].
- **Digital signatures**: the use of digital certificates for signing data transmission to the corresponding endpoint. The receiving side verifies the incoming data using the peer's public key, i.e., $SKEYID = prf(N_i|N_r, g^{xy})$.
- **Public key encryption**: the public/pair key can be used to encrypt the data instead of verifying digital signatures. It follows that
  $$SKEYID = prf(hash(N_i|N_r), CKY_i|CKY_r).$$
- **External authentication**: the external authentication server or services (such as Kerberos) can be used to authenticate end-points.

where N_x is the nonce payload of the initiator and responder, g^{xy} is the DH shared secret, CKY are 64-bit cookie values, and prf is the keyed pseudo-random (hash) function. The size of the key needed for authentication or encryption depends on the

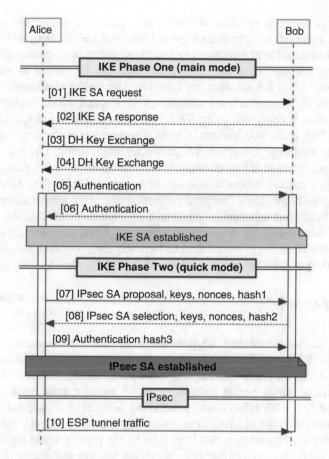

Fig. 2.15 Illustration of the IKE and IPsec connection establishment process. Messages marked with red arrows ([06]–[09]) are encrypted

type of used algorithm. However, the $SKEYID$ value is usually divided into three portions:

- $SKEYID_a$: for authentication of IKE phase 1
- $SKEYID_e$: for encryption of IKE messages [03] and [04]
- $SKEYID_d$: as input for establishment of IPsec SA in IKE phase 2

The third pair of messages performs authentication of endpoints to exchange their identities and generated hash values. It is interesting to note that any of the mentioned pairs can contain a vendor ID which identify the sender's characteristics and preferences, which in turn can be used to overcome NAT-related issues.

Aggressive mode is optional, and it was proposed as a means to accelerate IKE operations. Instead of exchanging three pairs of messages, it uses three messages as follows: negotiation and key exchange are performed in the first two messages, and the third message provides authentication. However, unlike the main mode, aggressive mode can use pre-shared secret keys for authentication, which creates a weak point susceptible to pre-shared key cracking and man-in-the-middle attacks.[3]

Upon completing phase one, IKE phase two can be started to establish the unidirectional IPsec SA sessions used for integrity and/or confidentiality of data traffic. IPsec SA sessions[4] are established in a *quick* mode under which all communication is encrypted using $SKEYID_e$ and an algorithm specified in IKE phase one. IKE phase two consists of three messages where the first two messages are used to exchange the IPsec SA parameter suggestions, DH values, nonces and a unique message ID to distinguish between multiple quick node messages. All messages contain authentication values based on $SKEYID_a$ to validate the exchange parameters (data integrity protection and data source authentication), while the final third message is introduced as liveness proof that the previously exchanged messages were successfully processed. Upon successfully exchanging messages, the IPsec SA is established to form the IPsec key material (KEYMAT), and depending on whether or not we demand Perfect Forward Secrecy (PFS):

$$KEYMAT = \begin{cases} prf(SKEYID_d, protocol \mid SPI \mid Ni_b \mid Nr_b) & \text{, without PFS} \\ prf(SKEYID_d, g^{xy} \mid protocol \mid SPI \mid Ni_b \mid Nr_b) & \text{, with PFS} \end{cases}$$

(2.1)

Details of the established SA are stored in Security Association Database (SAD). The SAD contains details of the algorithms based on pattern-matching of various fields in the IP header and rules stored in the database. Its purpose is to determine which packet belongs to which security association (SA). SAD contains the following information: source and destination IP address, SPI, IPsec security protocol (ESP or AH), IPsec mode (transport or tunnel), encryption or an authentication algorithm, secret keys used by the cryptographic algorithms, key length, SA lifetime, sequence number, anti-replay window and details of the type of traffic (such as port number or other) on which the specified SA is to be applied. In practice, inbound traffic filtering is based on the destination IP address, SPI value and IPsec security protocol. These three values can uniquely identify the SA and are used to perform lookup in the SAD for inbound traffic. If no SA is located, the incoming traffic is discarded.

The SA contains information about security actions that should be used to process the communication, but it does not fully specify which types of traffic should be protected. More detailed information about the type of traffic to be protected and under which conditions is stored in the Security Policy Database (SPD). The SPD contains details of the source and destination IP address, IP protocol (TCP, UDP

---

[3] Note that the pre-shared secret keys which are based on IP addresses can be used in main mode. However, because of this dependency, if pre-shared values are required, it is recommended that main mode be avoided [36].

[4] Multiple quick nodes establishing IPsec SA sessions can be performed under the protection of a single previously established IKE SA.

or other), port numbers (optional), IPsec protection to be applied (protect, bypass or discard) and a pointer to the SA record in the SAD. While SA entries in the SAD result from IKE operations, SPD details are usually defined using Command Line Interface (CLI). To check packet protection for outbound traffic, the SAD is searched according to IP or higher layer values. The identified SPD entry will point to an SA which identifies the entry in SAD. IPsec SA is a simplex (one-way) connection. IPsec, therefore, generally works in terms of an "SA bundle" connection consisting of two SAs, one in each direction to enable duplex (two-way) communication (Fig. 2.15).

Both IKE and IPsec SA have limited validity specified with the SA lifetime parameter. The lifetime threshold can be based on duration or the amount of network traffic. The SA cannot be extended, that is, the *rekeying* process is necessary to define new SA sessions before the SA expires.[5] In practice, the lifetime threshold is defined through the balance of safety and overhead. Setting the lifetime threshold to a lower value means greater security but also a greater amount of traffic exchanged for the establishment/rekeying of SA sessions. Additionally, as this establishment is based on authentication methods, it may interfere with the user's frequent request for manual authentication in host-to-gateway mode. RFC 4308 defines the upper bound for the lifetime as 24 hours for IKE SA and 8 hours for IPsec SA. In everyday use, it is recommended that smaller values such as 4 hours for IPsec SA and a few hours more for IKE SA are set [44, 45].

IKEv2 was proposed in 2005 (RFC 7296) to simplify and increase efficiency and strength against DoS [46]. In total, four messages in request-response form are used to establish IKE and IPsec SA. To create an additional pair of IPsec SA, only one additional message pair is needed per pair of SAs. In the initial pair of messages, both end-parties negotiate the set of cryptographic algorithms, exchange DH values and define a shared secret that will be used to derive additional keys to protect subsequent SAs. The second pair of messages is secured using a previously defined secret key and defined cryptographic algorithms. Both end-parties exchange identity and authentication check values. After the lifetime values expires, an additional "child" SA session is established, but without exchanging authentication values. Rekeying of the child SA session can include details of the cryptographic algorithm and new DH key exchange. IKEv2 also includes additional information queries between entities for use in collecting control information, i.e., to delete the SA.

2.4.4 IPsec and QKD

The first reported implementation of QKD keys with the IPsec framework was performed in the DARPA QKD network in 2002 [27]. The generated QKD keys

---

[5] The rekeying process can establish a new SA before (soft-rekeying) or after (hard-rekeying) the SA lifetime has expired.

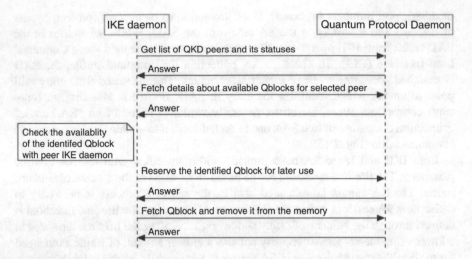

Fig. 2.16 Illustration of IKE and QPD daemon communication. Identification, availability check and reservation of Qblock keys is necessary to avoid conflicts before keys are used

were organized in Qblocks and stored in a local pool for each QKD peer. The IKE daemon fetched Qblocks from a pool as needed to create peer-to-peer session, i.e., IKE sessions were established only with a direct peer with which keys were successfully exchanged (Fig. 2.16).

The DARPA team noted that QKD and IKE authentication are redundant. That is, only one authentication is sufficient. They also noted that one method of generating SKEYID is to use pre-shared key such as QKD bits. However, since IKE phase one is used only to exchange the control-data needed to establish IKE phase two, no additional protection using QKD keys was needed [47].

Instead, they focused on the use of QKD bits to implement the Quantum Perfect Forward Secrecy (QPFS) concept. RFC 2409 defines the PFS concept which restricts use of the key to derive any additional keys [40]. In the sense of QPFS, it means that QKD bits are used for double-ephemeral DH exchange[6] for a limited session. The DARPA team therefore proposed replacing the DH values (g^{xy} in Eq. (2.1)) with QKD key bits for the hash which determines the IKE phase 2 keying material ($KEYMAT$). The Single-Photon Detector (SPD) and SAD databases are extended to distinguish SAs with PFS or QPFS support and with an SA session lifetime of up to 1 minute [48]. In this manner, traffic security can be differentiated using either a combination of DH and QKD keys or only DH or QKD bits to generate hash values. The need for efficient management was also noted to answer questions

---

[6] With Ephemeral Diffie-Hellman (DHE), a different seed is used for each session, and any leakage of the key still means that all previous communication sessions were secure. The DHE approach defines the PFS concept.

concerning the choice and policy definition of IPsec sessions, the policy of selecting QKD bits, and others.

Concerns were noted regarding the anticipated limitations in the duration of the IKE process. Usually, IKE phase one should be completed within 10 seconds, while phase two should be completed within a few seconds. Given that QKD key establishment can last longer, the risk of not having enough keys in the pool to respond to queries for the Qblock was observed. Nevertheless, increasing the timeouts is also a security concern. If the IKE daemon has to wait for the keys, an attacker has much more time to isolate and block IKE messages to interrupt communication. Since dedicated key management entities were not implemented, it is possible that the generated keys may not be entirely symmetrical if the key authentication step is excluded as a result of a reduction in time. This could lead to a cascading problem in the next steps when such a key is used to set up a new SA session or to encrypt/decrypt traffic over the active SA session.

In addition to the use of a symmetrical cipher with key material derived from QKD (e.g., Advanced Encryption Standard (AES)), the DARPA team reported the use of OTP with IPsec [48]. This approach is recommended for highly sensitive data and is based on modification of the IKE protocol to negotiate the use of one-time pads seeded with QKD bits. In this case, no need arises for IV values or padding bits that form an integral part of the ESP header. It is important to note the possible application of the IPComp protocol for data compression. Content compression can significantly reduce the consumption of QKD bits, but its efficiency depends on the type of transmitted data. Additionally, this implies the need to synchronize the QKD material since the entire reserved Qblock may be excessive for OTP encryption of compressed packets.

The DARPA network proposed the use of BBN Key Relaying (Fig. 2.4) for key transport.[7] If no Qblocks between distant nodes are available, a randomly generated Qblock will be relayed instead of message n. To minimize dependence on the established path through which key transport is accomplished, the authors proposed the use of several separate paths. In this variant, the original node A, shown in Fig. 2.4, sends reservation requests via all identified routes to the destination D. The XOR operations on Qblocks from all other nodes are then achieved on the destination node, which finally delivers only the final Qblock of all the collected values to the source node A. Because the risk exists that some of the nodes may be compromised during the key transport process, the authors suggested that the Qblock contain metadata which explicitly records the IDs of all nodes involved in key transport. In this way, if knowledge of a compromised node is obtained, all Qblocks executed through this node could be identified and be omitted.

Under the SECOQCproject, several approaches for the use of QKD keys in IPsec VPN connections were proposed. The Swiss team reported Secure Quantum Key Exchange Internet Protocol (SeQKEIP) [49]. Instead of modifications to the

---

[7] In DARPA documents, the mechanism of sharing the key between distant nodes is referred to as key transport [27].

IKE protocol which may violate standardization constraints, a combination of the ISAKMP framework with the new protocol was proposed. The SeQKEIP key exchange protocol implements three phases. The exchange of cryptographic keys using the QKD protocol is indicated by phase zero. In contrast to IKE, phase one does not include the exchange of selected DH groups or authentication based on signatures or digital certificates. Authentication is performed with the QKD protocol and use of pre-shared QKD keys. No cryptography keys are generated in this phase. The main aim of phase one is for end-point entities to exchange information about the encryption and authentication algorithms. In phase two, messages are encrypted and authenticated using the keys and algorithms agreed during the previous two phases. The main goal is to arrange an IPsec SA session as it is done in IKE phase two. The QKD key can be used to replace the IPsec session key ($KEYMAT$) or as an input to OTP. In the former case, the session key has a short duration, and its lifetime is equal to the time needed to establish a new session key using QKD. In the latter case, if sufficient QKD keys are available, each packet is encrypted using OTP. The authors expressed concern regarding the time required to establish a common session key given the long generation time of the applied QKD systems. Two solutions were therefore proposed: the use of traditional cryptographic algorithms in the IPsec tunnel but which are based on a common session key, and the use of session keys to directly encrypt data traffic resulting in low throughput.

One of the first key management solutions was presented in [50]. Specifically, it is about providing a network layer for QKD key material for different consuming applications, such as IPsec sessions. The key manager works according to a master/slave principle, serves key requests from consumers and negotiates requested queries with its slave key manager. The pool of keys which is established on both sides is served to authenticate communication between key managers and reserve the consumer's application.

In the context of the IPsec connection, the authors identified the quantum IKE (QIKE) protocol as a potential application for addressing the key manager with a request to reserve and serve keys. The authors noted the possibilities of performing authentication based on pre-shared keys ($SKEYID$), which in this context is performed using QKD keys (except for the initial phase).

In QIKE phase one, a similar function to IKE is performed. The main difference is in the exchange of the QKD key rate and an application ID in addition to the cryptographic algorithm and key lifetime parameters. Communication starts by a consumer A who sends the application ID and the address of the destination's key manager. Consumer A then sends the reservation requests to the key manager specifying the maximum key length (maxlen), the desired key rate (rate), and a lifetime. The key manager verifies the requests with the peer's key manager and reserve keys if a positive response is obtained. The authors proposed a buffer threshold value to raise an alert to reserved pools that will soon run out of keys. The buffer filling priority is proposed by first filling the local buffers necessary for traffic authentication between key managers and then filling the buffers with active threshold alarms.

To establish IPsec SAs, an array of SA proposals are exchanged between consumers in QIKE phase two. However, these proposals contain additional parameters such as the QKD key rate,[8] application ID and the maximum supported number of SAs. The specified key rate is the rate previously requested from the key manager to establish separate pools for each direction at the customer's end. The main aim of QIKE is to establish multiple SAs and provide high-speed IPsec implementation. The lifetime of SAs is specified in bytes to avoid synchronization problems. Upon expiry of the SA session, QIKE uses ISAKMP informational messages to delete the expired SAs from the SAD database. The expired SAs are replaced with the new SAs, which are instantly activated. To avoid synchronization problems in the case when deleted messages are not successfully delivered, increasing SPI values per connection are used. Although this approach may be vulnerable to DoS attacks, as discussed in [51], the authors argued that because of the specificity of the hardware implementation, the incremental approach cannot compromise the system. QIKE was one of the first solutions which involved establishing multiple IPsec sessions with the goal of maintaining high-speed communication. As shown in Fig. 2.17, QIKE is also a forerunner of some of the standardization approaches based on key manager entities. These approaches are discussed in the following chapters.

A similar solution was reported by the MagiQ technologies company through a patent issued in 2006 [52]. The authors identified standardization IPsec constraints which limit the frequency of key changes. To avoid these limitations, the authors proposed using the IKE protocol without modifications but applying the XOR operation on classic and QKD keys. The resulting keys are stored in SA tables on both sides of the VPN tunnel, and a number of SA connections are implemented (the authors suggested 2^{16} of 512 bytes, resulting in 32 Mbytes of required memory). Instead of rekeying, the next SA from the SA table is used, while the previously used SA is temporarily stored in memory to process any delayed packets in the network.

All the previously listed solutions rely on ISAKMP and IKEv1, whereas one of the first QKD solutions based on IKEv2 was presented in [53, 54]. The authors proposed the introduction of additional headers in the IKEv2 negotiations with the aim of exchanging information on the QKD key and the fallback mechanism. Specifically, the QKD KeyID header is transmitted within the initialization message IKE_SA_INIT; the header contains the following fields: payload length, QKD device ID, key length and key ID. The QKD device ID and key ID should uniquely identify the key in case the node is equipped with multiple QKD devices. The authors proposed the use of a transform field in the SA payload to specify the use of identified QKD keys (direct use or password-authenticated key). As shown in Fig. 2.18, the IKE_SA_INIT [01] messages carry the QKD KeyID identifier.

---

[8] The reason for the exchange of this parameter is not precisely specified by the authors. Consumers should have the same amount of key reserved after QIKE phase one. However, the additional exchange of QKD key rate can be used to verify the previously defined key rate, i.e., to avoid collision or misunderstanding of the amount of reserved keys.

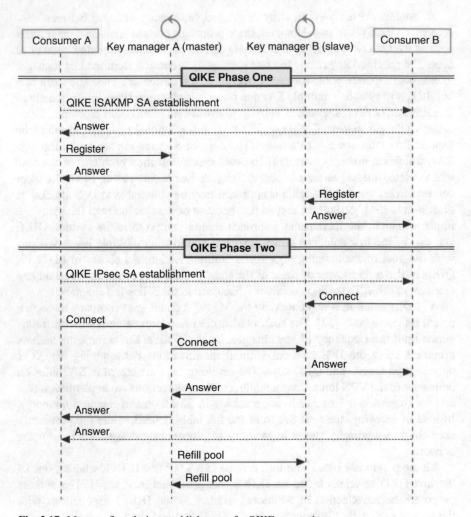

Fig. 2.17 Message flow during establishment of a QIKE connection

The standard Ni and KEi values used for DH key exchange are omitted. The HDR is the IKE header, while the SAi1 is a payload which includes the details of the cryptographic algorithms and transform field for QKD.

The responder answers with message [02] confirming the selected KeyID. The IKE_AUTH messages are encrypted using SA, whose key is chosen in IKE_SA_INIT. Messages [03] and [04] identify the QKD fallback mechanism, which can have one of the following values:

- WAIT_QKD: the system needs to wait for QKD devices to establish a new key
- DIFFIE-HELLMAN: indicates the use of DH keys within IKE in the existing IKE_SA

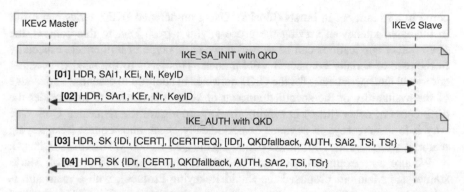

Fig. 2.18 Message flow during the establishment of an IKEv2 session using QKD keys [53]

- CONTINUE: indicates use of the most recent key until a new key becomes available

The specified QKD fallback mechanism is used when rekeying[9] is performed. The issue of a fallback mechanism can have an impact on other SA sessions. Namely, if the WAIT_QKD option is selected, which indicates waiting for available QKD keys, a precise notification mechanism which informs when keys become available is required. Additionally, the IKE daemon may be blocked because it is waiting to be served with new QKD keys, and it may delay the establishment or maintenance of other sessions which do not depend on the QKD key [55].

We can observe different approaches in the use of QKD keys from the examples listed above. They show that regardless of whether DH keys are combined or completely replaced with QKD keys, the IKE process still needs to generate random values which form an integral part of the security infrastructure. These random values ensure the lifespan of negotiations, protect against replay attacks, and generate nonce values within the authentication process, etc.

It is also important to look at approaches to using OTP with a combination of QKD keys. Although the authors of the examples listed above did not discuss the specifics of this approach (considering it is somewhat intuitive to understand), a few details require attention. Secret keys are usually stored in the IPsec SA database. IKE is in charge of filling the SA base with fresh keys and it is necessary to consider the efficiency of premature or on-demand charging. Initially, IKE often does not have complete information about the amount of keys needed for the lifetime of an SA session. Obtaining a large amount of keys in advance can result in unnecessary loss of those keys because they will never be used. Another problem is the storage of keys if larger quantities are involved. Nonetheless, QKD keys do not

---

[9] Due to the lack of a description of the IKA_SA_INIT procedure without a QKD key, it follows that IKE_SA_INIT must be performed with the QKD key whereas rekeying can be accomplished using DIFFIE-HELLMAN.

come in a stream but in bursts (blocks). Using on-demand IKE queries can result in increased latency in serving the process with a fresh key. In this context, the operation of applications can be slowed down at this type of bottleneck, especially in the case of serially connected VPN tunnels. This leads to the need to consider additional QoS parameters for the QKD network. In addition to essential knowledge of the availability of the specified amount of keys, it is necessary to consider the differences in delay (jitter) in serving the keys to ensure the requested QoS. One way is to reserve keys in advance or periodically check and request fresh keying material, which are the approaches defined in the QoS models we discuss in Chap. 3.

We also note examples not based on modification of the IKE protocol. Marksteiner and Maurhart proposed the Rapid Rekeying Protocol, whose main aim is robustness and swiftness in rekeying (key synchronization) [51, 56]. Instead of modifying the IKE protocol to perform key establishment, the authors noted that the QKD protocol is in charge of this task. Thus, the IPsec ESP tunnel can be established using manual keying (linux command setkey[10]). In the case of automated fast rekeying, there is a need to synchronize the large number of established SA sessions. Additionally, the authors noted that the IPsec SPI value is transmitted unencrypted and that in its simplest form it can be a simple incremental integer value. However, this type of implementation can reveal details of the IPsec rekeying rates to an attacker. The authors suggested hashing the SPI value using the SHA-1 algorithm over the previous SPI value. The old SPI value is XORed with 32 bits of QKD bits which represent salt value. The SHA-1 hash is calculated over the XORed value. The proposed algorithm uses the leftmost 32 bits of the SHA-1 output to define a new SPI value. Salts and the first SPI value can be taken from the QKD key, representing an acceptable consumption of key material even for low performance QKD systems.

To provide efficient synchronization, the authors proposed the implementation of a dedicated IPsec AH transport channel. All key change commands are exchanged over this channel authenticated but not encrypted. The SPIs and keys of this channel are changed periodically to ensure higher security. The control channel key change mechanism implements a three-way handshake consisting of a key change request and acknowledge and re-acknowledge messages. Failure of key change in the control channel will result in the repetition of messages until a timeout is reached to stop and prevent all further communication. A number of active SAs can exist simultaneously on the slave side, allowing the decryption of delayed packets. However, to limit the number of active SAs, the oldest SA is deleted on the slave before acknowledgment to the master is provided.

The refresh rate of the keys depends primarily on the availability of keys. The validity period of the key duration can be calculated from $P_k = 2 \cdot kQ$, where Q is the amount of keys usually dictated by the key rate, and k is the length of the IPsec specified key multiplied by two for two IPsec unidirectional tunnels [51]. For

---

[10] The authors used the *Netlink* network socket-oriented protocol to provide a key feed for the IPsec SAs in the Linux kernel.

example, for QKD systems with a key rate of 100 kbps and an AES key length of 128 bits used in IPsec SA, $P_k = 2.56$ ms.

The refresh rate can be also limited by the amount of traffic protected. To provide ITS communication, it is necessary to supply an amount of keys identical to the amount of data to be encrypted with OTP. If authentication is performed in addition to encryption, then more keys than the amount of data is required, which is challenging to provide given the capabilities of current generation QKD systems.

2.4.5   Passive and Active Eavesdropping

The eavesdropping mechanism basically assumes a natural rate of errors which appears unnaturally high in the presence of an eavesdropper. An implementation of the basic concept yields tolerable error rates that QKD typically assumes. These error rates have recently been subject to criticism [57], various problems being outlined in how the tolerable bounds were computed. Independently of these criticisms, proposed improvements in detecting eavesdroppers by using a finer-grained classification of where errors arise [58] assume an adversary's access to (quantum) memory [59] and rely on non-orthogonal [60], non-binary [61] or more complex interleaved encoding [62]. One of the adversarial strategies for eavesdropping is the exploitation of imperfect sources (emitting an inaccurate number of photons [63, 64] or backflashes/backreflections [65, 66]), hardware trojans [67, 68], attempts at quantum cloning [69, 70], or the use of laser-pulses sent back into the device to either detect or change (even damage) its internal configuration [71–73]. Side-channel vulnerabilities are a different yet no less successful means of attack [74, 75] and may include hiding an adversary's presence by deceiving the detection mechanism [76]. Perhaps the most intuitive yet technologically "cheap" attack is transforming the eavesdropping facility of QKD against itself: the notion here is to simply disrupt communication with a DoS mounted by passive eavesdropping [77] or to exploit the trusted node model itself by using eavesdropping to logically cut paths (by letting them run dry) to redirect traffic flow through nodes which are easy to infiltrate [78]. These attacks target classic software vulnerabilities in the firmware of the quantum device (it is, after all, in most cases only a conventional computer controlling the device, shipped with classic software insecurities and imperfections). Countermeasures against such attacks involve secure software design (being outside the scope of this book), proper key management and arrangement of local key stores [79] or optimized yet randomized source routing [80–82]. The latter in turn requires a proper and compatible network design: a problem which has received interest [5] yet is known to be computationally intractable in the most general case [83, 84]. Other attacks bypass eavesdropping by launching the authenticity mechanisms behind QKD to replace the legitimate receiver of the qubit stream and spare eavesdropping at all, which is essentially a person-in-the-middle attack. One such successful attack on a QKD system, although exploiting physical imperfections of the implementation rather than relying on breaching the authentication key, is reported in [76]. Liu et al. [85] described an attack which exploits a form of QKD that is vulnerable only if finitely many keys are created. Attacks exploiting certain implementations of logical steps, such as iterated sifting [86] or quantum secret sharing techniques, are given by Scarani and Gisin [87].

The reported history of vulnerabilities in QKD can broadly be classified in (1) attacks exploiting imperfections in the devices themselves, and (2) attempts to invalidate the assumptions beneath security proofs related to QKD. Neither form of attack is new or unique to quantum devices since software vulnerabilities are known to carry higher risk (they are easier to exploit) than attacks by cryptanalysis. The latter also rests on assumptions whose violation in practice occasionally happens (for example, [88]). The overall risk posed by attacks to QKD depends on which vulnerability is the cheapest, i.e., the most economical, to exploit. With strong proofs of security along the line and highly developed and reliable devices, an attacker may simply choose the weakest spot to hit a system, which in many cases may be a non-quantum component (software running on devices, inadequately secured secret stores from which the QKD key can be extracted through a "conventional" side-channel attack, etc.).

2.4.6 QoS Constraints in QKD Network

Summarizing the above, the basic requirements for QoS in QKD networks can be divided into two groups:

1. **QoS constraints defined by the IP network**: affect data exchange performance and communication performance when exchanging key information (such as IKE rekeying and other control messages). This group includes:

   - Bandwidth on demand
   - End-to-end delay
   - Delay variation (jitter)
   - Error/loss rate without retransmission

2. **QoS constraints defined by the QKD entities**: parameters which define the possibility of establishing secure communication and its quality. This group includes:

   - Amount of available key material, also referred to as "key bandwidth"
   - The delay variations in providing QKD keys

   Note that the listed parameters are not necessarily the only which can be applied, i.e., users can restrict the type of devices from which they want to collect QKD keys. Namely, different devices practice different techniques for generating keys, which can be reflected in the epsilon security of the key. To be precise, an adversary cannot gain any information about the secret key, except with the probability epsilon [89]. In a similar manner, the specific route for key relaying can also be defined, thus limiting the number of hops in the communication. Another parameter is the key validity time, which allows users to limit the period of use of the key to counter threats of its misuse. Yet all of these requirements are application specific, and for high-quality and secure communication in QKD networks, it is necessary to find a compromise.

2.5 Similarities Between QKD and Ad Hoc Networking

The requirements described above can significantly limit the space for implementing a QoS solution. However, if the characteristics of QKD networks are observed from a high level, similarities can be seen in ad hoc networks where similar restrictions apply [90, 91].

Below we list the basic specifications of the QKD network:

- QKD link is always implemented in a point-to-point manner and is characterized by two basic features: limited length, and the key rate, which is inversely proportional to the length [25]. Additionally, QKD link is active or available only when sufficient key material can be provided through that link. Viewed in this context, we may notice similarities with WiFi wireless links of limited range where communication performance depends on the distance of the user from the antenna.
- QKD networks do not currently implement quantum amplifiers/repeaters which can amplify the signal over a greater distance. Therefore, communication most often occurs in a hop-by-hop approach with key-relay support [92].

If the characteristics of the Mobile Ad Hoc Network (MANET) network are considered in the same manner, the similarities with QKD technology may be quickly noticed. The MANET network consists of battery-powered mobile nodes. Therefore, special attention is given to energy efficient QoS solutions. Additionally, nodes are connected by a decentralized, self-organizing structure without hierarchical superior nodes performing network management and control. As a result of their mobility, WiFi links executed in a MANET network can often lead to unstable routes [93].

The amount of energy available in a MANET node's battery can be linked to the amount of keys in a QKD node.[11] Restrictions on the range of WiFi links can be related to the restrictions on the range of quantum channels. Additionally, the lack of a dedicated network infrastructure in MANET networks and the execution of communication in a hop-by-hop manner can be related to the lack of quantum repeaters and implementation of hop-by-hop key transfer. From all the above, we

---

[11] We emphasize here that there is a difference in the context that one battery is used to operate an antenna which sends and receives signals from multiple paths whereas QKD keys are defined only on a single specific link.

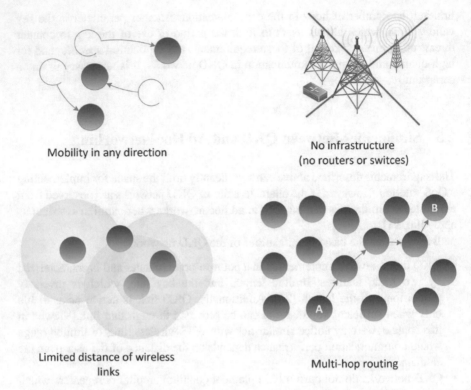

Fig. 2.19 Similarities between MANET and QKD

can conclude that solutions proven to be effective in MANET networks are also interesting for application in QKD networks.

Although at first glance MANET and QKD networks have nothing in common, a simple analysis of the features of these networks reveals their similarities (Fig. 2.19). However, what clearly distinguishes these two networks is their purpose. MANET networks are designed to provide fast and easy communication in critical situations where no dedicated network infrastructure is available (such as battlefield scenarios, search and rescue operations during natural disasters, and others). The basic task of QKD technology is to improve the level of security, more precisely, to enable ITS secure communication. This difference in purpose may limit the application of certain solutions because they may be suitable for some technologies yet contrary to the requirements of others. As a basic example, we can cite the manner in which routing protocols are implemented. In MANET networks, proactive routing protocols are often implemented by flooding the network with queries of the path to a specific destination. In QKD networks, such solutions may be undesirable. Namely, a QKD network is based on a trusted-relay approach which requires the security and trustworthiness of all QKD nodes. Given the nature of QKD links, an attacker will identify nodes as critical points instead of attacking the links themselves. Specifically, nodes which do not participate in the path should not even be aware that key transfers are taking place on another part of the network. If an attacker possesses this information, he might launch an attack on routing protocols with the goal of redirecting communication to nodes under his control. Therefore, routing information should be encrypted or at least authenticated [94, 95]. More details on routing in QKD networks are provided in Chap. 6. All of the above leads to the conclusion that network solutions based on network flooding may not be suitable for application in QKD networks.

2.6 Summary

This chapter provided an overview of key distribution in trusted relay QKD networks. In addition to a review of basic QoS parameters widely known in IP networks, the chapter also listed QoS parameters related to QKD networks. In this context, it is important to point out that QKD keys can be integrated into existing security frameworks such as IPsec in several ways. These methods differ in their criteria for integrating QKD keys according to refresh rate, the number of established SA tunnels and other management aspects. However, their settings largely depend on the type and amount of data sought to be protected and the level of security required. In addition, the recognized similarity between QKD and ad-hoc networks suggests that new solutions for QKD system integration could be identified using already existing and well-accepted approaches. In this context, and knowing the targeted QoS parameters of interest, network administrators can select the QoS mechanisms necessary for organizing their QoS architectures and provide a satisfactory level of security and communication quality.

References

1. ITU-T. (1994). Recommendation E. 800, terms and definitions related to quality of service and network performance including dependability. Technical Report.
2. Bailey, P. (2002). *Deploying CISCO voice over IP solutions*. Indianapolis: Cisco Press.
3. Lu, G. (2000). Issues and technologies for supporting multimedia communications over the internet. *Computer Communications, 23*(14–15), 1323–1335. ISSN 01403664. https://doi.org/10.1016/S0140-3664(00)00179-1
4. Park, K. I. (2005). *QoS in packet networks*. ISBN 0-387-23389-X. https://doi.org/10.1007/b101424
5. Alleaume, R., Roueff, F., Diamanti, E., & Lutkenhaus, N. (2009). Topological optimization of quantum key distribution networks. *New Journal of Physics, 11*(7), 75002. ISSN 1367-2630. https://doi.org/10.1088/1367-2630/11/7/075002
6. Salvail, L., Peev, M., Diamanti, E., Alléaume, R., Lütkenhaus, N., & Länger, T. (2010). Security of trusted repeater quantum key distribution networks. *Journal of Computer Security, 18*(1), 61–87. ISSN 18758924. https://doi.org/10.3233/JCS-2010-0373

7. ETSI ISG QKD. (2019). Quantum key distribution (QKD); device and communication channel parameters for QKD deployment. 1, 1–19. https://www.etsi.org/deliver/etsi_gs/QKD/001_099/012/01.01.01_60/gs_qkd012v010101p.pdf

8. Liao, S. K., Lin, J., Ren, J. G., Liu, W. Y., Qiang, J., Yin, J., Li, Y., Shen, Q., Zhang, L., Liang, X. F., Yong, H. L., Li, F. Z., Yin, Y. Y., Cao, Y., Cai, W. Q., Zhang, W. Z., Jia, J. J., Wu, J. C., Chen, X. W., et al. (2017). Space-to-ground quantum key distribution using a small-sized payload on tiangong-2 space lab. *Chinese Physics Letters, 34*(9), 1–6. ISSN 17413540. https://doi.org/10.1088/0256-307X/34/9/090302.

9. Schmitt-Manderbach, T., Weier, H., & Fürst, M. (2007). Experimental demonstration of free-space decoy-state quantum key distribution over 144 km. *Physical Review Letters, 98*, 010504.

10. Nauerth, S., Moll, F., Rau, M., Fuchs, C., Horwath, J., Frick, S., & Weinfurter, H. (2013). Air-to-ground quantum communication. *Nature Photonics, 7*(5), 382–386. ISSN 1749-4885. https://doi.org/10.1038/nphoton.2013.46

11. Wang, J. Y., Yang, B., Liao, S. K., Zhang, L., Shen, Q., Hu, X. F., Wu, J. C., Yang, S. J., Jiang, H., Tang, Y. L., Zhong, B., Liang, H., Liu, W. Y., Hu, Y. H., Huang, Y. M., Qi, B., Ren, J. G., Pan, G. S., Yin, J., et al. (2013). Direct and full-scale experimental verifications towards ground-satellite quantum key distribution. *Nature Photonics, 7*(5), 387–393. ISSN 17494885. https://doi.org/10.1038/nphoton.2013.89

12. Arnon, S., & Kupferman, J. (2019). *Effects of weather on drone to IoT QKD. Lecture notes in computer science* (vol. 11527). Berlin: Springer. ISBN 978-3-03020-950-6. https://doi.org/10.1007/978-3-030-20951-3_5

13. Liao, S. K., Cai, W. Q., Handsteiner, J., Liu, B., Yin, J., Zhang, L., Rauch, D., Fink, M., Ren, J. G., Liu, W. Y., Li, Y., Shen, Q., Cao, Y., Li, F. Z., Wang, J. F., Huang, Y. M., Deng, L., Xi, T., Ma, L., et al. Satellite-relayed intercontinental quantum network. *Physical Review Letters, 120*(3), 30501. ISSN 10797114. https://doi.org/10.1103/PhysRevLett.120.030501

14. Calderaro, L., Agnesi, C., Dequal, D., Vedovato, F., Schiavon, M., Santamato, A., Luceri, V., Bianco, G., Vallone, G., & Villoresi, P. (2019). Towards quantum communication from global navigation satellite system. *Quantum Science and Technology, 4*(1), 1–8. ISSN 20589565. https://doi.org/10.1088/2058-9565/aaefd4

15. Kerstel, E., Gardelein, A., Barthelemy, M., Fink, M., Joshi, S. K., & Ursin, R. (2018). Nanobob: A CubeSat mission concept for quantum communication experiments in an uplink configuration. *EPJ Quantum Technology, 5*(1), 1–30. ISSN 21960763. https://doi.org/10.1140/epjqt/s40507-018-0070-7

16. Aleksic, S., Winkler, D., Franzl, G., Poppe, A., Schrenk, B., & Hipp, F. (2013). Quantum key distribution over optical access networks. In *Proceedings of the 2013 18th European Conference on Network and Optical Communications & 2013 8th Conference on Optical Cabling and Infrastructure (NOC-OC&I)* (pp. 11–18). https://doi.org/10.1109/NOC-OCI.2013.6582861

17. Wang, L.-J., Zou, K.-H., Sun, W., Mao, Y., Zhu, Y.-X., Yin, H.-L., Chen, Q., Zhao, Y., Zhang, F., Chen, T.-Y., & Pan, J.-W. (2017). Long-distance copropagation of quantum key distribution and terabit classical optical data channels. *Physical Review A, 95*(1), 012301 (2017). ISSN 2469-9926. https://doi.org/10.1103/PhysRevA.95.012301

18. Kumar, R., Qin, H., & Alléaume, R. (2015). Coexistence of continuous variable QKD with intense DWDM classical channels. *New Journal of Physics, 17*, 043027. ISSN 13672630. https://doi.org/10.1088/1367-2630/17/4/043027

19. Karinou, F., Brunner, H. H., Fung, C. H. F., Comandar, L. C., Bettelli, S., Hillerkuss, D., Kuschnerov, M., Mikroulis, S., Wang, D., Xie, C., Peev, M., & Poppe, A. (2018). Toward the integration of CV quantum key distribution in deployed optical networks. *IEEE Photonics Technology Letters, 30*(7), 650–653. ISSN 10411135. https://doi.org/10.1109/LPT.2018.2810334

20. Mao, Y., Wang, B.-X., Zhao, C., Wang, G., Wang, R., Wang, H., Zhou, F., Nie, J., Chen, Q., Zhao, Y., Zhang, Q., Zhang, J., Chen, T.-Y., & Pan, J.-W. (2018) Integrating quantum key distribution with classical communications in backbone fiber network. *Optics Express, 26*(5), 6010. ISSN 1094-4087. https://doi.org/10.1364/OE.26.006010

21. Sergienko, A. V. A. V. (2005). *Quantum communications and cryptography* (vol. 2005). Boca Raton: CRC Press. ISBN 978-0-84933-684-3.
22. Lucamarini, M., Yuan, Z. L., Dynes, J. F., & Shields, A. J. (2018). Overcoming the rate–distance limit of quantum key distribution without quantum repeaters. *Nature, 557*(7705), 400–403. ISSN 1476-4687. https://doi.org/10.1038/s41586-018-0066-6
23. Yuan, Z., Plews, A., Takahashi, R., Doi, K., Tam, W., Sharpe, A., Dixon, A., Lavelle, E., Dynes, J., Murakami, A., Kujiraoka, M., Lucamarini, M., Tanizawa, Y., Sato, H., & Shields, A. J. (2018). 10-Mb/s quantum key distribution. *Journal of Lightwave Technology, 36*(16), 3427–3433. ISSN 07338724. https://doi.org/10.1109/JLT.2018.2843136
24. Zhang, Y., Chen, Z., Pirandola, S., Wang, X., & Zhou, C. (2020). Long-distance continuous-variable quantum key distribution over 202. 81 km of fiber. *Physical Review Letters, 125*(1), 10502. ISSN 1079-7114. https://doi.org/10.1103/PhysRevLett.125.010502
25. Kollmitzer, C., & Pivk, M. (2010). *Applied quantum cryptography* (vol. 797). Berlin: Springer. ISBN 364-2-04829-3. https://doi.org/10.1007/978-3-642-04831-9
26. Hayashi, M., Iwama, K., Nishimura, H., Raymond, R., & Yamashita, S. (2007). Quantum network coding. In W. Thomas & P. Weil (Eds.), *Lecture Notes in Computer Science. Lecture Notes in Computer Science* (vol. 4393, pp. 610–621). Berlin: Springer. ISBN 978-3-540-70917-6. https://doi.org/10.1007/978-3-540-70918-3
27. Elliott, C., & Yeh, H. (2007). DARPA quantum network testbed. Technical Report, BBN Technologies Cambridge, New York.
28. Peev, M., Pacher, C., Alléaume, R., Barreiro, C., Bouda, J., Boxleitner, W., Debuisschert, T., Diamanti, E., Dianati, M., Dynes, J. F., Fasel, S., Fossier, S., Fürst, M., Gautier, J.-D., Gay, O., Gisin, N., Grangier, P., Happe, A., Hasani, Y., et al. (2009). The SECOQC quantum key distribution network in Vienna. *New Journal of Physics, 11*(7), 075001. ISSN 1367-2630. https://doi.org/10.1088/1367-2630/11/7/075001
29. Sasaki, M. (2011). Tokyo QKD network and the evolution to secure photonic network. In *CLEO:2011 - Laser Applications to Photonic Applications* (vol. 1, p. JTuC1). Washington: OSA. ISBN 978-1-55752-910-7. https://doi.org/10.1364/CLEO_AT.2011.JTuC1
30. Mirza, A., & Petruccione, F. (2010). Realizing long-term quantum cryptography. *Journal of the Optical Society of America B, 27*(6), A185. ISSN 0740-3224. https://doi.org/10.1364/JOSAB.27.00A185
31. Shimizu, K., Honjo, T., Fujiwara, M., Ito, T., Tamaki, K., Miki, S., Yamashita, T., Terai, H., Wang, Z., & Sasaki, M. (2014). Performance of long-distance quantum key distribution over 90-km optical links installed in a field environment of Tokyo metropolitan area. *Journal of Lightwave Technology, 32*(1), 141–151. ISSN 0733-8724. https://doi.org/10.1109/JLT.2013.2291391
32. Mehic, M., Niemiec, M., Rass, S., Peev, M., Aguado, A., Martin, V., Schauer, S., Poppe, A., Pacher, C., & Voznak, M. (2020). Quantum key distribution: A networking perspective. *ACM Computing Surveys, 53*(5), 96.
33. Boyd, C., Mathuria, A., & Stebila, D. (2020). *Protocols for authentication and key establishment. Information security and cryptography*. Berlin: Springer. ISBN 978-3-662-58145-2. https://doi.org/10.1007/978-3-662-58146-9
34. Atkinson, R., & Kent, S. (1998). IP authentication header. RFC 2402. https://rfc-editor.org/rfc/rfc2402.txt
35. Wang, Y.-S., Eggert, L., & Touch, J. D. (2004). Use of IPsec transport mode for dynamic routing. RFC 3884. https://rfc-editor.org/rfc/rfc3884.txt
36. Doraswamy, N., & Harkins, D. (2003). *IPSec: The new security standard for the internet, intranets, and virtual private networks*. Upper Saddle River: Prentice Hall. ISBN 0-13-046189-X.
37. Voznak, M., Rezac, F., Halas, M., Bulucea, C. A., Kalamani, N., Mastorakis, N., & Mladenov, V. (2010). Speech quality evaluation in IPsec environment. In *12th International Conference on Networking, VLSI and Signal Processing, Stevens Point World Scientific and Engineering Academy and Society (WSEAS)* (vol. 12, pp. 49–53). ISBN 978-9-60474-162-5.

38. Monsour, B., Stevens, W. R., Pereira, R., & Shacham, A. (2001). IP payload compression protocol (IPComp). RFC 3173. https://rfc-editor.org/rfc/rfc3173.txt

39. Sataloff, R. T., Johns, M. M., & Kost, K. M. (2014). *Network security, firewalls, and VPNs*. Burlington: Jones & Bartlett Learning. ISBN 978-1-62623-977-7.

40. Carrel, D., & Harkins, D. (1998). The internet key exchange (IKE). RFC 2409. https://rfc-editor.org/rfc/rfc2409.txt

41. Krawczyk, H. (1996). SKEME: A versatile secure key exchange mechanism for Internet. In *Proceedings of Internet Society Symposium on Network and Distributed Systems Security* (pp. 114–127). Washington: IEEE Computer Society Press. ISBN 0-8186-7222-6. https://doi.org/10.1109/NDSS.1996.492418

42. Orman, H. (1998). The OAKLEY key determination protocol. RFC 2412. https://rfc-editor.org/rfc/rfc2412.txt

43. Goldreich, O., Goldwasser, S., & Micali, S. (1986). How to construct random functions. *Journal of the ACM, 33*(4), 792–807. https://doi.org/10.1145/6490.6503

44. Frankel, S., Kent, K., Lewkowski, R., Orebaugh, A. D., Richey, R. W., & Sharma, S. R. (2005). Guide to IPsec VPNs recommendations of the national institute (p. 126). Nist Special Publication.

45. Hoffman, P. E. (2005). Cryptographic suites for IPsec. RFC 4308. https://rfc-editor.org/rfc/rfc4308.txt

46. Kaufman, C., Hoffman, P. E., Nir, Y., Eronen, P., & Kivinen, T. (2014). Internet key exchange protocol version 2 (IKEv2). RFC 7296. https://rfc-editor.org/rfc/rfc7296.txt

47. Dervisevic, E., & Mehic, M. (2021). Overview of quantum key distribution technique within IPsec architecture. In A. Adrot, R. Grace, K. Moore, & C. Zobel (Eds.), *Proceedings of the 18th International Conference on Information Systems for Crisis Response and Management ISCRAM 2021* (pp. 1–10). New York: ACM.

48. Elliott, C., Pearson, D., & Troxel, G. (2003). Quantum cryptography in practice. In *Proceedings of the 2003 Conference on Applications, Technologies, Architectures, and Protocols for Computer Communications - SIGCOMM'03* (p. 227). https://doi.org/10.1145/863981.863982

49. Sfaxi, M. A., Ghernaouti-Hélie, S., Ribordy, G., & Gay, O. (2005). Using quantum key distribution within IPSEC to secure MAN communications. In *IFIP Open Conference on Metropolitan Area Networks Architecture, Protocols, Control, And Management, MAN*.

50. Neppach, A., Pfaffel-Janser, C., Wimberger, I., Loruenser, T., Meyenburg, M., Szekely, A., & Wolkerstorfer, J. (2008). Key management of quantum generated keys in IPSEC. In *SECRYPT 2008 - International Conference on Security and Cryptography, Proceedings* (pp. 177–183).

51. Marksteiner, S., & Maurhart, O. (2015). A protocol for synchronizing quantum-derived keys in IPsec and its implementation. In *ICQNM 2015: The Ninth International Conference on Quantum, Nano/Bio, and Micro Technologies, Venice* (vol. 9, pp. 35–40). ISBN 978-1-61208-431-2. https://doi.org/10.13140/RG.2.1.4756.4001

52. Berzanskis, A., Hakkarainen, H., Lee, K., & Hussain, M. R. (2006). Method of integrating QKD with IPSec. https://patents.google.com/patent/US7602919B2/en

53. Nagayama, S., & Van Meter, R. (2009). IKE for IPsec with QKD. Technical Report. https://tools.ietf.org/html/draft-nagayama-ipsecme-ipsec-with-qkd-01

54. Nagayama, S., & Van Meter, R. (2014). IKE for IPsec with QKD. Technical Report. https://datatracker.ietf.org/doc/html/draft-nagayama-ipsecme-ipsec-with-qkd-01

55. Mink, A., Frankel, S., & Perlner, R. (2009). Quantum key distribution (QKD) and commodity security protocols: Introduction and integration. *International Journal of Network Security & Its Applications, 1*(2), 101–112. ISSN 0975-2307. https://doi.org/10.5281/zenodo.1239668

56. Marksteiner, S. (2014). *An approach to securing IPsec with Quantum Key Distribution (QKD) using the AIT QKD software*. Ph.D. Thesis, CAMPUS 02 University of Applied Sciences, Graz, Austria.

57. Iwakoshi, T. (2017). On problems in security of quantum key distribution raised by Yuen. In Gruneisen, M. T., Dusek, M., & Rarity, J. G. (Eds.), *Quantum information science and technology III. Proceedings of SPIE* (p. 3). Bellingham: SPIE. ISBN 978-1-51061-348-5. https://doi.org/10.1117/12.2278625

58. Khan, M. M., Xu, J., & Beige, A. (2011). Improved eavesdropping detection in quantum key distribution. arXiv preprint arXiv:1112.1110.
59. Bocquet, A., Leverrier, A., & Alléaume, R. (2011). Optimal eavesdropping on QKD without quantum memory. arXiv preprint arXiv:1106.0329.
60. Sasaki, T., Yamamoto, Y., & Koashi, M. (2014). Practical quantum key distribution protocol without monitoring signal disturbance. *Nature, 509*(7501), 475–478. https://doi.org/10.1038/nature13303
61. Banerjee, S., Maiti, B., & Saha, B. (2017). Analysis of eavesdropping in QKD with qutrit photon states. In Chaki, R., Saeed, K., Cortesi, A., & Chaki, N. (Eds.), *Advanced computing and systems for security: Volume three* (pp. 161–171). Singapore: Springer. ISBN 978-981-10-3409-1. https://doi.org/10.1007/978-981-10-3409-1_11
62. Lizama-Pérez, L., López, J., & de Carlos López, E. (2017). Quantum key distribution in the presence of the intercept-resend with faked states attack. *Entropy, 19*(1), 4. ISSN 1099-4300. https://doi.org/10.3390/e19010004
63. Félix, S., Gisin, N., Stefanov, A., & Zbinden, H. (2001). Faint laser quantum key distribution: Eavesdropping exploiting multiphoton pulses. *Journal of Modern Optics, 48*(13), 2009–2021. ISSN 0950-0340. https://doi.org/10.1080/09500340108240903
64. Sajeed, S., Radchenko, I., Kaiser, S., Bourgoin, J.-P., Pappa, A., Monat, L., Legré, M., & Makarov, V. (2015). Attacks exploiting deviation of mean photon number in quantum key distribution and coin tossing. *Physical Review A, 91*(3), 431. ISSN 1050-2947. https://doi.org/10.1103/PhysRevA.91.032326
65. Pinheiro, P. V. P., Chaiwongkhot, P., Sajeed, S., Horn, R. T., Bourgoin, J.-P., Jennewein, T., Lütkenhaus, N., & Makarov, V. (2018). Eavesdropping and countermeasures for backflash side channel in quantum cryptography. *Optics Express, 26*(16), 21020–21032.
66. Jain, N., Anisimova, E., Khan, I., Makarov, V., Marquardt, C., & Leuchs, G. (2014). Trojan-horse attacks threaten the security of practical quantum cryptography. *New Journal of Physics, 16*(12), 123030. https://doi.org/10.1088/1367-2630/16/12/123030
67. Sajeed, S., Minshull, C., Jain, N., & Makarov, V. (2017). Invisible Trojan-horse attack. *Scientific Reports, 7*(1), 8403. https://doi.org/10.1038/s41598-017-08279-1
68. Jain, N., Stiller, B., Khan, I., Makarov, V., Marquardt, C., & Leuchs, G. (2015). Risk analysis of trojan-horse attacks on practical quantum key distribution systems. *IEEE Journal of Selected Topics in Quantum Electronics, 21*(3), 168–177. https://doi.org/10.1109/JSTQE.2014.2365585
69. Siomau, M., & Fritzsche, S. (2010). Efficiency of the eavesdropping in B92 QKD protocol with a QCM. In A. Sergienko, S. Pascazio, & P. Villoresi (Eds.), *Quantum Communication and Quantum Networking* (pp. 267–274). Berlin: Springer. ISBN 978-3-642-11731-2.
70. Okubo, Y., Buscemi, F., Tomita, A., & Lvovsky, A. (2009). Proposal of an eavesdropping experiment for BB84 QKD protocol with 1->3 phase-covariant quantum doner. In *AIP Conference Proceedings* (pp. 355–358). Woodbury: AIP. https://doi.org/10.1063/1.3131347
71. Lydersen, L., Wiechers, C., Wittmann, C., Elser, D., Skaar, J., & Makarov, V. (2010). Hacking commercial quantum cryptography systems by tailored bright illumination. *Nature Photonics, 4*(10), 686–689. https://doi.org/10.1038/nphoton.2010.214
72. Bugge, A. N., Sauge, S., Ghazali, A. M. M., Skaar, J., Lydersen, L., & Makarov, V. (2014). Laser damage helps the eavesdropper in quantum cryptography. *Physical Review Letters, 112*(7), 70503. https://doi.org/10.1103/PhysRevLett.112.070503
73. Tanner, M. G., Makarov, V., & Hadfield, R. H. (2014). Optimised quantum hacking of superconducting nanowire single-photon detectors. *Optics Express, 22*(6), 6734–6748. https://doi.org/10.1364/OE.22.006734
74. Sajeed, S., Huang, A., Sun, S., Xu, F., Makarov, V., & Curty, M. (2016). Insecurity of detector-device-independent quantum key distribution. *Physical Review Letters, 117*(25), 250505. https://doi.org/10.1103/PhysRevLett.117.250505
75. Makarov, V., Bourgoin, J.-P., Chaiwongkhot, P., Gagné, M., Jennewein, T., Kaiser, S., Kashyap, R., Legré, M., Minshull, C., & Sajeed, S. (2016). Creation of backdoors in quantum communications via laser damage. *Physical Review A, 94*(3), 325. ISSN 1050-2947. https://doi.org/10.1103/PhysRevA.94.030302

76. Gerhardt, I., Liu, Q., Lamas-Linares, A., Skaar, J., Kurtsiefer, C., & Makarov, V. (2011). Full-field implementation of a perfect eavesdropper on a quantum cryptography system. *Nature Communications, 2*(2027), 349. ISSN 2041-1723. https://doi.org/10.1038/ncomms1348

77. Cao, Z. (2013). Eavesdropping or disrupting a communication — On the weakness of quantum communications. Cryptology ePrint Archive, Report 2013/474.

78. Rass, S., & König, S. (2012). Turning quantum cryptography against itself: how to avoid indirect eavesdropping in quantum networks by passive and active adversaries. *International Journal On Advances in Systems and Measurements, 5*(1–2), 22–33.

79. Schartner, P., & Rass, S. (2009). How to overcome the 'Trusted Node Model' in quantum cryptography. In *2009 International Conference on Computational Science and Engineering, Los Alamitos, California* (pp. 259–262). Piscataway: IEEE. ISBN 978-1-4244-5334-4. https://doi.org/10.1109/CSE.2009.171

80. Schartner, P., & Rass, S. (2009). How to overcome the 'Trusted Node Model' in quantum cryptography. In *2th IEEE International Conference on Computational Science and Engineering, CSE 2009* (vol. 3, pp. 259–262). ISBN 978-0-76953-823-5. https://doi.org/10.1109/CSE.2009.171

81. Tanaka, A., Maeda, W., Takahashi, S., Tajima, A., & Tomita, A. (2008). Randomize technique for quantum key and key management system for use in QKD networks. In *SECOQC Demonstration Conference*.

82. Rass, S. (2013). On game-theoretic network security provisioning. *Springer Journal of Network and Systems Management, 21*(1), 47–64. https://doi.org/10.1007/s10922-012-9229-1

83. Rass, S. (2014). Complexity of network design for private communication and the P-vs-NP question. *International Journal of Advanced Computer Science and Applications, 5*(2), 148–157.

84. Rass, S., Kollmitzer, C., 2010. Adaptive Cascade, in: Applied quantum cryptography. Springer, Lecture Notes in Physics (LNP, volume 797), pp. 49–69, ISBN: 978-3-642-04831-9.

85. Liu, X., Zhang, B., Wang, J., Tang, C., Zhao, J., & Zhang, S. (2014). Eavesdropping on counterfactual quantum key distribution with finite resources. *Physical Review A, 90*(2), 022318. ISSN 1050-2947. https://doi.org/10.1103/PhysRevA.90.022318

86. Tamaki, K., Lo, H.-K., Mizutani, A., Kato, G., Lim, C. C. W., Azuma, K., & Curty, M. (2018). Security of quantum key distribution with iterative sifting. *Quantum Science and Technology, 3*(1), 14002.

87. Scarani, V., & Gisin, N. (2001). Quantum key distribution between N partners: Optimal eavesdropping and Bell's inequalities. *Physical Review A, 65*(1), 661. ISSN 1050-2947. https://doi.org/10.1103/PhysRevA.65.012311

88. Lenstra, A. K., Hughes, J. P., Augier, M., Bos, J. W., Kleinjung, T., & Wachter, C. (2012). Ron was wrong, Whit is right. Cryptology ePrint Archive, Report 2012/064.

89. Renner, R. (2005). *Security of quantum key distribution*. Ph.D. Thesis, Swiss Federal Institute of Technology Zurich. http://arxiv.org/abs/quant-ph/0512258

90. Fazio, P., De Rango, F., & Sottile, C. (2016). A predictive cross-layered interference management in a multichannel MAC with reactive routing in VANET. *IEEE Transactions on Mobile Computing, 15*(8), 1850–1862. ISSN 1536-1233. https://doi.org/10.1109/TMC.2015.2465384

91. Fazio, P., Tropea, M., De Rango, F., & Voznak, M. (2016). Pattern prediction and passive bandwidth management for hand-over optimization in QoS cellular networks with vehicular mobility. *IEEE Transactions on Mobile Computing, 1233*(c), 1. ISSN 1536-1233. https://doi.org/10.1109/TMC.2016.2516996

92. Alleaume, R., Branciard, C., Bouda, J., Debuisschert, T., Dianati, M., Gisin, N., Godfrey, M., Grangier, P., Langer, T., Lutkenhaus, N., Länger, T., Lütkenhaus, N., Monyk, C., Painchault, P., Peev, M., Poppe, A., Pornin, T., Rarity, J., Renner, R., et al. (2014). Using quantum key distribution for cryptographic purposes: A survey. *Theoretical Computer Science, 560*(P1), 62–81. ISSN 03043975. https://doi.org/10.1016/j.tcs.2014.09.018

93. Sarkar, K., Basavaraju, T. G., & Puttamadappa, C. (2008). *Ad Hoc mobile wireless networks* (vol. 1). Boca Raton: CRC Press. ISBN 084-9-39567-4.

94. Maurhart, O., Lorunser, T., Langer, T., Pacher, C., Peev, M., & Poppe, A. (2009). Node modules and protocols for the Quantum-Back-Bone of a quantum-key-distribution network. In *2009 35th European Conference on Optical Communication* (pp. 3–4).
95. Mehic, M., Fazio, P., Rass, S., Maurhart, O., Peev, M., Poppe, A., Rozhon, J., Niemiec, M., & Voznak, M. (2020b). A novel approach to quality-of-service provisioning in trusted relay quantum key distribution networks. *IEEE/ACM Transactions on Networking, 28*(1), 168–181. ISSN 1063-6692. https://doi.org/10.1109/TNET.2019.2956079

Chapter 3
Quality of Service Architectures of Quantum Key Distribution Networks

Quality of Service (QoS) architecture models the structure and methods of applying QoS mechanisms to satisfy QoS objectives. It is often referred to the QoS model because it uniquely defines which QoS mechanisms (such as those listed in Sect. 2.3) are used and how they are implemented. QoS architecture also defines network traffic processing policies and the choice of signaling and routing protocols.

Today's Internet is able to interconnect multiple network domains and may be considered a macro-system which provides domain-to-domain data forwarding, with end-to-end QoS delivery guarantees. Although the best-effort approach widely dominates and all QoS mechanisms are layered on top of the existing Internet infrastructure, two frameworks have become the main architectures for providing Internet QoS [1].

Integrated Services (IntServ) is a QoS framework with the capability of reserving resources dynamically for single flows (it is also referred to as "per-flow based") [2, 3]. In this framework, each router needs to provide the exact amount of resources to satisfy QoS constraints for the specific traffic flows. Two additional service classes with respect to the existing best-effort Internet model are provided: guaranteed and controlled load services. The first provides an upper bound for the end-to-end queuing delay and is dedicated to applications with strict real-time requirements, with a guaranteed bandwidth and no queuing packet losses [4]. Clearly, each router needs to be informed of the flow characteristics in terms of the amount of resources to be reserved. The controlled load service class is designed to share a given amount of bandwidth between different flows while ensuring a service similar to the best-effort, when network use is low (packet loss and delay are generally maintained below defined bounds) [5]. By the concept of per-flow reservation, the IntServ architecture is able to deliver QoS guarantees. However, the introduction of state information inside each router represents a fundamental change to the current Internet architecture, and also a major issue when the approach is applied to backhaul connections. In fact, when a single device should manage a large number of flows, it is challenging to maintain a separate queue for each flow state because of

© Springer Nature Switzerland AG 2022
M. Mehic et al., *Quantum Key Distribution Networks*,
https://doi.org/10.1007/978-3-031-06608-5_3

scalability issues. For these reasons, IntServ does not appear to be adequate for inter-domain communications, while it is suitable for local domains with high-bandwidth applications [6].

Differentiated Services (DiffServ) [7, 8] is the second QoS framework proposed to address some of the issues arising from the deployment of IntServ. In contrast with the previous framework, it is referred as "per-aggregate-class based" architecture, because it applies the features of packet marking [9] by modifying some bits in the packet header to indicate that the marked packet belongs to a particular QoS level or QoS class. To do so, DiffServ pushes the computational complexity to the boundary of the network (to the edge routers, for example), in the sense that given the relative small number of connections, it can perform packet classification more easily than locally (a core router inside an AS, for example). In this manner, scalability is attained by leaving the marking operations to the boundaries of the network. DiffServ provides two service models for enhancing the "best-effort" approach: Premium [10], for which a peak rate is guaranteed (with very low queuing delay), and Assured [11], for which statistical provisioning is considered. In the second case, two priority classes are defined, and packets are marked with one of them according to their service profile; thus, lower priority packets have a greater probability of being dropped.

Both frameworks must be managed adequately to implement what they provide in terms of QoS. To this aim, Internet QoS is based on two main groups of mechanisms: data paths and control paths. The first is responsible for individual packet operations, such as classification, marking, policing and scheduling [12–15], while the second is aimed at configuration of the involved intermediary nodes (flow admission and resource reservation). One of the most important control path algorithms is Call Admission Control (CAC) [16–19]. It is a module implemented in each access node (an access point, a femtocell, a base station, etc.) which decides whether, on the basis of the characteristics and requests, a new flow should be admitted into the network, while continuing to satisfy all the QoS specifications of the previously admitted nodes. To manage the intra-domain resources (bandwidth management and reallocation) and inter-domain agreements (bandwidth negotiation and parameter configuration), a logical entity called a bandwidth broker (BB) is defined as a component of the control path algorithm. Finally, access to the network must be regulated under the given conditions with some policies dedicated to controlling the resources and services accessed by users and applications [20]. In the following sections, details of IntServ and DiffServ are discussed for a clearer understanding of how they function.

3.1 Integrated Services

IntServ is dedicated as support for real-time and non-real-time IP services. The extension to QoS is necessary to meet the growing requests for real-time applications for a very wide range of new applications, including tele-networking (e.g., telemedicine), conferencing, distributed computing, etc. Many other factors have also influenced the development of QoS requirements:

- Recent servers and workstations now come equipped with a range of built-in hardware (including audio and video codecs, multimedia devices and CPUs, etc.) which have become inexpensive;
- IP broadcasting, the Internet of Things, new generation networks and new architectures have spread around the world;
- Complex digital audio and video applications have been introduced into the market as a result of developments in mobile computing.

The development of the above-mentioned technologies have demonstrated that real-time applications often do not work well with the Internet best-effort paradigm because of variable queuing delays and congestion losses. In other words, the Internet infrastructure must be "modified" to support mainly real-time QoS, for which some control over end-to-end packet delays is needed.

IntServ architecture represents an Internet service model which integrates best-effort, real-time and controlled link sharing services. The requirements and mechanisms for IntServ have been the subject of much discussion and research. The model proposed in [2] includes two types of service targeted toward real-time traffic: guaranteed service and predictive service. It integrates these services with controlled link-sharing and is designed to work well with both multicast and unicast. Some assumptions are also made: resources must be explicitly allocated to satisfy application requirements (this implies that "resource reservation" and "admission control" are key operations in QoS for setting the required bandwidth). Given that resource availability is dynamic over time, one approach is to provide service guarantees using a priori reservations. Users must be able to obtain service requests whose quality is predictable (between some given ranges). It is therefore very important to specify the main features of the required traffic (bandwidth, delay, etc.). In this sense, intermediate routers should provide a mechanism for reservation of resources, to provide special QoS for specific user packet streams (flows). From this requirement, the concept of flow-specific state (and its specific setup in IntServ aware nodes) was derived, representing an important and fundamental change in the Internet model, where the concept of flow state has never been taken into account. This new approach preserves the intrinsic robustness of classic Internet mechanisms while allowing efficient management of network resources. To understand the enormous advantages introduced by IntServ, we have to recall the definition of "flow" as a distinguishable stream of related datagrams created by single user application and requiring the same QoS. Generally, all packets receive the same QoS and are typically forwarded using a strict FIFO queuing discipline.

All forwarders (routers) along the path must implement all the appropriate QoS mechanisms for each flow, in accordance with the service model. Thus, each device implements a so-called "traffic control" function to create different QoS. It is composed of three main components:

- The packet scheduler (able to extract packets from the queue in the required order and with the required guarantees);
- The classifier (able to assign the required priorities to packets);
- Admission control (able to evaluate whether a new flow can be served while maintaining the same QoS for already admitted flows).

All these components should be managed by a proper signaling and reservation protocol (also called a setup protocol) which is able to create and maintain a flow-specific state at the endpoints and all forwarding nodes. Generally, at the application level, the required (or desired) QoS is specified through the implementation of a list of parameters, called "flowspec".

The reservation protocol is able to carry the flowspec, which is passed to each admission control module for acceptability, and finally, used to configure a proper packet scheduling mechanism.

Figure 3.1 illustrates the internal structure of an IntServ-aware forwarding node. In the lower part, the forwarding sequence is illustrated: it is executed each time a packet needs to be sent and therefore should be highly optimized (generally it

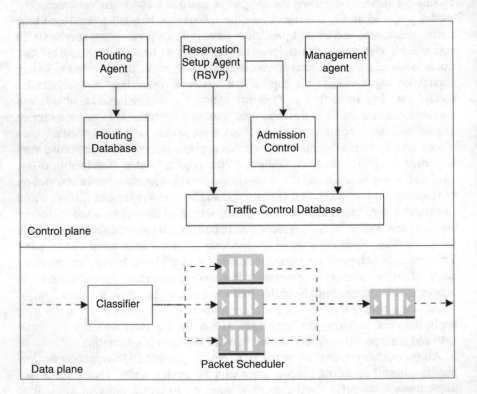

Fig. 3.1 Router structure in terms of reservation components for IntServ: two broad functional set of components are evident. The upper part of the figure illustrates the background running code, while the lower part shows the forwarding sequence

is hardware assisted). The input driver is responsible for interpreting the protocol header (such as the IP header), while the internet forwarder executes a "suite-dependent" classification procedure for passing the packet and its class to the output driver. The final one implements a type of packet scheduling.

The upper part represents the key components of the coding: the background code is stored in the node's memory and is normally executed by the CPU. In this way, the required data structures are created and the forwarding sequence (illustrated in the lower part of the figure) can be controlled. The routing agent software is configured to follow a particular routing protocol, based on the routing database (table, entries, etc.). The reservation setup agent implements the reservation protocol and sets up the resource reservation requests for the admission control module: if the request is accepted, then the appropriate changes are made to the classifier and packet scheduler database to implement the desired QoS. The final software component is the network management agent: it directly modifies the classifier and packet scheduler databases to configure and implement the controlled link-sharing and admission control algorithms.

When referring to a QoS-aware host implementation, the structure is the same as the one depicted in Fig. 3.1, apart from the addition of the application layer: it is responsible for generating the data to be transmitted and receiving the data destined for the host as the destination node. Obviously, an application which needs a real-time QoS for a flow must invoke a local reservation setup agent: the host may use a dedicated Application Programmers Interface (API) to invoke a resource setup, or the reservation request may be executed directly from the operating system. One of the key metrics in IntServ networks is packet delay. The upper and lower delay bounds are requested from the application layer of the source host. Some applications may need to receive the data contained in each packet by a certain time (real-time applications) or may tolerate some wasted waiting times (elastic applications).

To manage QoS adequately in an IntServ architecture, it is necessary to define the "reservation model", which is able to describe how an application should negotiate in requesting and dealing with a QoS level. In the first scenario, the application may refer to the CAC algorithm so that the network accepts or refuses the request, but it is also considered to refer to multiple levels of flow specification (flowspec). That is, the application asks for a particular QoS level, but the network may be able to grant a lower resource level, informing the application of what QoS has actually been granted. The second scenario may provide a completely different behavior: each sender propagates the offered flowspec through a multicast-tree so that each involved router can adjust the received flowspec to the effectively available resource. In this way, each receiver knows what type of QoS is available into the network, and it can then generate the requested flowspec (on the basis of the received offer), which is propagated back toward the sender. The second scenario is also known as "two-pass" reservation scheme. It offers the possibility to know the allowed delay "in-advance". In conclusion, the type of services available in IntServ can be summarized as follows:

- Guaranteed delay bounds: some theoretical results ([21, 22]) show that if a QoS-aware router implements Weighted Fair Queue (WFQ) scheduling [23] and if the traffic can be characterized by a model with some bounds, such as the token bucket, then an upper bound of the overall network delay can be obtained;
- Link sharing: the same WFQ scheme can provide controlled link sharing. In this case, the focus is not on bounding the delay, but on limiting the overload shares on a link, while allowing any mix of traffic to proceed if spare capacity is available. This feature is already available in commercial routers;
- Predictive real-time services: it is more suitable than guaranteed service, in that it provides a bound on the delay with negligible jitter, and therefore the receiver can have an estimation. It should be underlined that WFQ leads to a guaranteed bound, but it is not guaranteed to be low enough. This is because traffic packets are not mixed into a single queue but are separated (larger delay).

3.1.1   RSVP Protocol

To give the nodes the opportunity to communicate with the network and express the desired level of QoS, Interv is based on a signaling mechanism. In this way, each end-system can send packets with information destined for another end-system through the network's intermediate devices. The most used is Resource Reservation Protocol (RSVP). Referring to Fig. 3.2, we notice that RSVP is a signaling protocol and invokes specific functions defined in the node (generally a router) or the network segment which delivers QoS. Obviously, RSVP is not aware of the semantics contained in the information it carries.

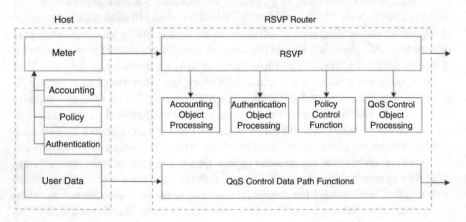

Fig. 3.2 High level architecture for IntServ

We infer from Fig. 3.2 that RSVP does not provide any functions, only communication services. The IntServ model is largely based on the RSVP signaling protocol for signaling QoS purposes. Because of its dependence on RSVP, the IntServ model is often referred to as the RSVP model.

The RSVP is a transport-level signaling protocol specified in [24, 25]. It implements separate queues on routers for each network flow. Its dependence on current buffer management is also its main obstacle, which is reflected in its limited scalability. To address this limitation to some extent, RSVP implements soft-state reservations which require a periodic refresh to avoid the reservation time-out. In light of the IntServ approach, RSVP uses two types of flowspecs to inform path nodes about QoS reservation requirements:

- Traffic Specifications (T-Spec) uses token bucket model specifications to depict traffic characteristics

  - Bucket rate r (bps),
  - Peak rate p (bps),
  - Bucket depth b (bits),
  - The maximum size of packet which can be accepted M (bits),
  - The minimum policed unit which specifies the minimum size of the packet. Any packet with a smaller size is counted as the minimum policed unit m (bits).

- Request Specification (R-Spec) carries details of the requested amount of bandwidth to be reserved in Guaranteed Service type.

  - Service rate R (bps) denoting the requested amount of bandwidth to be reserved for a specified flow,
  - Slack term S (μs) denoting the additional tolerated amount of delay considering the total end-to-end delay requirement.

The T-Spec is used to determine the R value in the downstream procedure (from source to destination). Based on the T-Spec values, the token bucket model (r,p,b,M) can calculate the requested bandwidth R and buffer space B for each accepted flow. The guaranteed service type assumes that:

1. A theoretical *fluid-flow traffic model* can be used. That is, the source node generates traffic according to a variable bit-rate continuous-time process $\rho(t)$.
2. One token corresponds to sending one bit of data traffic.
3. Initially, the token bucket is filled with a total of b tokens. The buffer is empty.
4. $p > R > r$, where R is the service rate allocated to specific traffic flow using RSVP reservation.
5. The propagation delays and packet sizes are neglected ($m=0$; $M=0$).

From this idealistic model (Fig. 3.3), we can calculate the maximum amount of traffic which can be served in the T_b. From assumption 3) from the previous list, the bucket is filled with b tokens which can serve peak traffic at the peak rate p. Therefore, the maximum burst size (MBS) which can be expected from the token

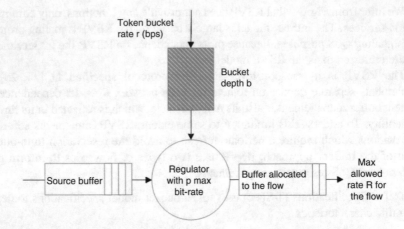

Fig. 3.3 Illustration of the token bucket (r, p, b) model

bucket shape can be calculated as follows:

$$MBS = T_b \cdot p = b + r \cdot T_b \tag{3.1}$$

It follows that the time interval of maximum service is:

$$T_b = \frac{b}{p - r} \tag{3.2}$$

After T_b, the output rate is defined with the arrival rate of new tokens r. To obtain the equation which defines the maximum delay expected from such a system, we use the network-calculus tool. The arrival rate indicating the total number of bits generated up to time t in a fluid model can be denoted $\alpha(t)$. That is, this non-decreasing arrival function of time is:

$$\alpha(t) = \int_0^t \rho(t)dt \tag{3.3}$$

If a real traffic source is considered, the arrival rate $\alpha(t)$ is the limiting value for the cumulative traffic $A(\tau, \tau + t)$ generated in the interval $[\tau, \tau + t]$, i.e., $A(\tau, \tau + t) \leqslant \alpha(t)$.

According to Eq. (3.1) and as shown in Fig. 3.4, the arrival curve $\alpha(t)$ is regulated by the token bucket shape as follows:

$$\alpha(t) = \min\{pt, b + rt\} = \begin{cases} pt, & t < \frac{b}{p-r} \\ b + rt, & t \geqslant \frac{b}{p-r} \end{cases} \tag{3.4}$$

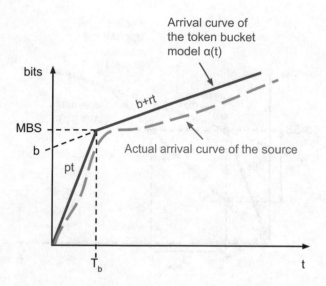

Fig. 3.4 Illustration of the token bucket model (r, p, b) regulation of the arrival curve

In a similar manner, the departure curve $\beta(t)$ defines the number of bits departing the node up to time t. Assuming that the node's maximum output/service rate $R(R > r)$ which characterizes the serving curve $\sigma(t)$ is defined by the model and assuming that $R = const$, the departure curve from the node can be defined as follows:

$$\beta(t) = \min\{\alpha(t), \sigma(t)\}, t > 0 \tag{3.5}$$

That is, after consuming b tokens from the bucket, the incoming curve $\alpha(t)$ is limited by the slope of $b + rt$. The output curve from system $\beta(t)$ is limited by the number of arriving bits, even though the maximum output rate R is given.

Figure 3.5 denotes the maximum occupancy of the transmission buffer B_{max} and the maximum experienced delay D_{max} for the considered flow. The occupancy of the transmission buffer at a generic time t can be calculated as:

$$B(t) = \alpha(t) - \beta(t) \tag{3.6}$$

As shown in Fig. 3.5, the maximum delay D_{max} can be calculated as follows:

$$D_{max} = t' - T_b = \frac{b}{R}\left(\frac{p - R}{p - r}\right) \leq \frac{b}{R} \tag{3.7}$$

and is bounded by the ratio b/R in an idealistic fluid model. Similarly, the maximal buffer size B_{max} is bounded by the bucket b and can be calculated as:

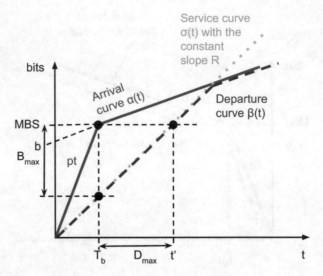

Fig. 3.5 Illustration of the departure curve $\beta(t)$ as a function of time t

$$B_{max} = p \cdot T_b - R \cdot T_b = b\left(\frac{p-R}{p-r}\right) \leq b \tag{3.8}$$

If a realistic case is considered in which the maximum packet size M and the overall system delay T_0 are taken into account, the behavior of the token bucket model (r, p, b, M) is analyzed. In that case, the incoming curves start from the value of M at the moment $t = 0$, assuming that $b > M, T > 0$ and $T_0 > 0$. Although now the arrival curve is described using T-Spec parameters:

$$\alpha(t) = \min\{pt + M, b + rt\} \tag{3.9}$$

it has no significant effect on the manner of calculating the values of D_{max} and B_{max}. Specifically, D_{max} can be calculated as follows:

$$D_{max} = \frac{b-M}{R} \cdot \left(\frac{p-R}{p-r}\right) + \frac{M}{R} + T_0 \tag{3.10}$$

If the approach described above is to be applied to specific IntServ GS specifications, token bucket T-spec items (r, p, b, M) which characterize the network flow should be considered. Specifically, the allocation of bandwidth R and the definition of buffer capacity B on all nodes of the route from source to destination are required, all with the provision of a defined maximum delay Δ_{max}.

This reservation is made during the RSVP setup phase. Specifically, the PATH message generated by the source node is updated by each node on the path to the destination. The PATH message contains fixed T-Spec specifications but also advertisement specification (ADSPEC) settings which accumulate on the path. ADSPEC contains two parameters which describe the deviations from the idealistic fluid-flow traffic mode. The rate-dependent correction term C accounts the serialization and fragmentation delay, while the rate-independent delay term D includes propagation and service delays. These two values are accumulated along the path, resulting in the updated calculation of end-to-end delay D_{max}. Let $C_{tot} = \sum_{i=1}^{N} C_i$ and $D_{tot} = \sum_{i=1}^{N} D_i$. Then:

$$D_{max} = \begin{cases} \frac{M+C_{tot}}{R} + D_{tot}, & R \geq p \geq r \\ \frac{b-M}{R} \cdot \left(\frac{p-R}{p-r}\right) + \frac{M+C_{tot}}{R} + D_{tot}, & p \geq R > r \end{cases} \tag{3.11}$$

Upon receiving summarized values, the destination node needs to find the minimum value $R_{min} \geq r$ which considers the requested delay bound D_{obj} using Eq. (3.12). If the denominator of (3.12) is considered, we notice that the R_{min} value exists only when $D_{obj} > D_{tot}$.

$$R_{min} = \begin{cases} \frac{M+C_{tot}}{D_{obj}-D_{tot}}, & R \geq p \geq r \\ \frac{p \cdot \frac{b-M}{p-R}+M+C_{tot}}{D_{obj}-D_{tot}+\frac{b-M}{p-r}}, & p \geq R > r \end{cases} \tag{3.12}$$

Consider the example of a path with four nodes where the destination node requires a total delay of $D_{obj} = 600$ ms. Based on the collected values, the destination will calculate $R_{min} \sim 609$ kbps. As shown in Fig. 3.6, the RESV messages with calculated R-Spec are sent upstream to the source node along the same path. Since the intermediate node i does not know C_{tot} or D_{tot} nor the D_{obj} value, it can only check whether it is possible to reserve $R_i = R_{min}$ to the specific flow. The intermediate node i will verify that the sum of previously reserved rates and requested rate R_{min} is less than the scheduler total rate R. If the reservation is accepted, the RSVP message is passed upstream.

We may note that R-Spec includes an additional parameter denoted *slack term*. It is used when the targeted value D_{obj} is larger than the bound value of D_{tot}. Since the destination node calculates the minimal value R_{min}, there might be additional values R which satisfy the Eq. (3.12). The slack term $S = D_{obj} - D_{max}$ (ms) is calculated and passed to the upstream node in the R-Spec field. Since the intermediate node i does not know the summarized delay nor the objective D_{obj} values, it can use the value S to increase its internal delay. The slack value denotes the amount by which the D_{tot} delay will be below the D_{obj}, assuming each router along the path reserves R_{min} bandwidth. Exchange of S values provides greater flexibility for the individual routers in performing local reservations and can increase the likelihood of successful end-to-end reservation. However, in some cases it can increase the probability of not accepting the reservation request when a longer path is considered.

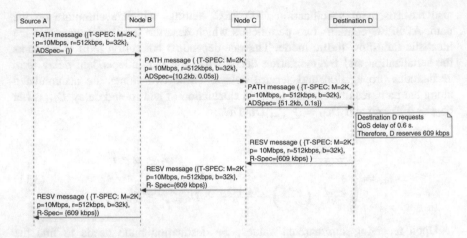

Fig. 3.6 RSVP reservation setup process, exchange of PATH and RSVP messages

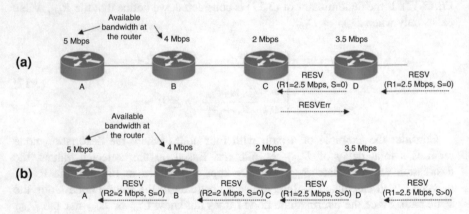

Fig. 3.7 RSVP response to PATH request message: (**a**) Reservation request denied at the intermediate router C (slack term = 0); (**b**) Reservation request accepted after use of the slack term value at the intermediate router

As shown in Fig. 3.7a, the RESV reservation message is declined at the intermediate node because of a lack of available bandwidth. Introduction of the slack value allows an intermediate router to adopt to the reservation request and forward the RESV message further along the path (Fig. 3.7b).

3.1.2   ETSI 004: QKD Application Interface

At the time of writing (summer 2021), several standards consider the integratation of QKD keys into IP networks. Here, particular emphasis is given to the ETSI 004 and ETSI 014 standards, which specify API between applications and QKD network components [26].

Both of these standards assume that the QKD network is partitioned into domain-restricted areas where the basic and central point for providing QKD keys to end-user applications is performed by Key Manager System (KMS), or simply, key managers (KM). Different types of application may request QKD keys from the KMS. Optical switches, encryption modules, security management systems and other end-user applications can be viewed as an Secure Application Entity (SAE). However, the starting point for collecting and processing requests from SAE applications is KMS. Its responsibility is also to manage the keys generated by QKD devices securely and to enable communication between remote SAE entities upon request.

In the context of the QoS model, the ETSI 004 standard is distinct because it includes QoS specifications within the several queries. It follows a session-oriented approach whereby sessions are uniquely identified using a Key_Stream_ID (KSID) value. A session expired due to inactivity can be reactivated (if its settings are not deleted from the KMS memory) using the same KSID value in OPEN_CONNECT query. The ETSI 004 standard defines three API functions:

- OPEN_CONNECT—a function for establishing a key stream session and reserving keys in accordance with QoS-defined specifications. The input values conveyed by this message are source and destination SAE identifiers and QoS specifications. The response is KSID and status values.
- GET_KEY—a function for obtaining previously reserved keys. The input values conveyed by this message are the KSID identifier, optional metadata and key positioning index (discussed further below). The response is a message containing the stream of keys, optional metadata and output status which reports on the success of the request.
- CLOSE—a function to close the key stream session and release previously reserved resources. The input value conveyed by this message is the KSID identifier, resulting in the status message which reports on the success of the request.

Illustrated in Fig. 3.8, Alice's SAE application wants to contact Bob's SAE application and contacts the nearest KMS system by sending the OPEN_CONNECT message. This message attempts to establish the reservation of QKD keys on both sides of the future communication path with the potential definition of the minimum required QoS conditions. This call is blocked until a connection is established or the timeout specified for establishing a connection (timeout) in the QoS-defined specifications expires.

After receiving the OPEN_CONNECT request, the KMS will consider fulfilling the request and contact the remote KMS to make the reservation. It is important to note here that neither ETSI 004 nor ETSI 014 defines the communication between KMS entities (blue lines in Fig. 3.8). Therefore, a KMS can wait for a response from a remote KMS before making a reservation, or in the simplest case, it can respond to SAE without waiting for a response from a remote KMS. Note that in the first

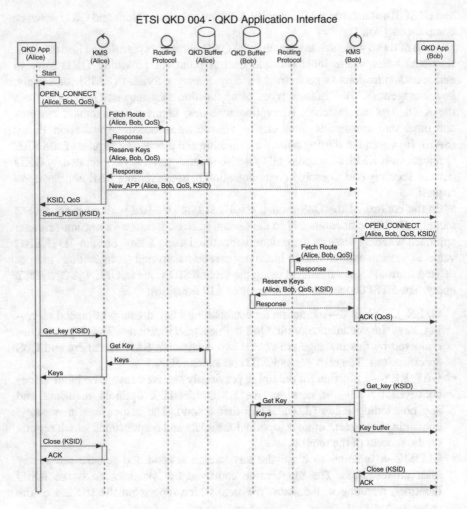

Fig. 3.8 Sequence diagram of an ETSI 004 Application Interface exchanging QoS specifications. The blue dotted lines are outside the scope of the ETSI 004 standard [26]

case, waiting for a response may result in increased delay, while in the second case, a collision may occur as a result of insufficient synchronization of the KMS entities (KMS on Alice's side claims to have enough resources to accept the reservation request, while the KMS on Bob's side reports no available resources). This question is especially significant if a chain of multiple KMSs of the path is considered.

We may notice a similarity between the ETSI 004 approach and the IntServ architecture. Namely, KMS is now placed in the position of CAC, which decides whether meeting QoS requirements is possible.

As noted in [27], QKD key storage can be identified as a token bucket traffic shaping mechanism. As shown in Fig. 3.9, the volume of available key material

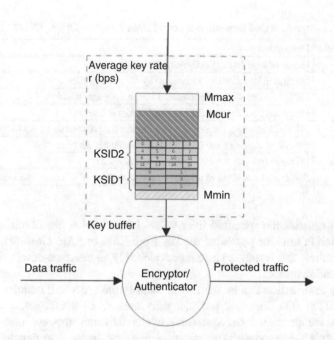

Fig. 3.9 QKD token bucket model. The key buffer is used to store cryptographic keys whose average generation rate is denoted r (bps). M_{max} indicates the capacity of the buffer, the M_{min} indicates the initial number of pre-shared keys for first rounds of post-processing, and M_{cur} denotes the amount of key material at the time of measurement. KSID1 and KSID2 denote the reserved key material of different chunk sizes for multiple sessions following the ETSI 004 approach

depends on the stored amount of key material M_{cur} in the buffer and the key generation rate r. Assuming the rate r is constant in time when the QKD devices are installed, the KMS can calculate whether new queries can be served and with which QoS specifications.

A list of QoS parameters specified in ETSI 004 messages is shown in Table 3.1. ETSI 004 proposes the organization of keys in logical buffers where the size of each key is denoted by the *key chunk size* value. The algorithms for key synchronization and the key merge/shrink option are not considered in the standard, and it should be implemented as a part of KMS–KMS communication. Having logical key buffer reserved, the SEA application will fetch a key using the GET_KEY command, denoting the index position of the key in the logical buffer (logical buffer for KSID1 and KSID2 are respectively shown in orange and blue in Fig. 3.9).

We also note that SAE does not define the explicit value of the maximum number of keys which can be reserved. Instead, the maximum and minimum key rate is provided, allowing KMS to calculate the availability of keys based on the key generation rate r and the available number of keys in the key buffer M_{cur}. In contrast to the specification of jitter discussed in Sect. 2.4.6, ETSI 004 defines jitter as the maximum expected deviation in key delivery measured in bps. That is, if KMS is not able to deliver a key of the specified key chunk size, it can deliver a shorter

Table 3.1 Description of QoS parameters specified in the ETSI 004 OPEN_CONNECT function

QoS parameter	Description
Key chunk size	length of keys (bytes) reserved in the key buffer
Max bps	Maximum key rate (bps) expected by the SAE application
Min bps	Minimum key rate (bps) tolerated by the SAE application
Jitter	Permitted flow delay variation (bps) in key delivery
Priority	Information about the priority of a request for processing by KMS
Timeout	Permitted waiting time (ms) to obtain responses for requests
Time to live	Lifespan (sec) of the established KSID session
Metadata	Additional metadata of requests, such as number of intermediate hops and others

key which satisfies the specified jitter value. In Sect. 2.4.6, we described jitter as the deviation in time for providing the key from KMS to SAE. However, ETSI 004 does not address this question since it provides only an interface description, not an implementation thereof.

Priority information can be marked to identify the order and manner of serving requests. ETSI 004 does not provide guidelines about queuing techniques, but it denotes the permitted timeout (ms) which indicates the acceptable waiting time to obtain the response. The metadata field can be used to denote additional specifications, such as the age of the key (the key generation time stamp from which the key can be considered valid), the number of hops that are visited to deliver the key in a key relay approach, and others.

The standard acknowledges that KMS-KMS communication is required to perform synchronization, stating that *"the QKD manager is responsible for reserving and synchronizing the keys at the two ends of the QKD link through communication with its peers"*. However, without a clear definition of the same, the process of obtaining keys can lead to a collision in the handling of requests. Consider the GET_KEY queries shown in Fig. 3.8. Due to the lack of feedback on whether reservation was successfully made, SAE Alice can request the key using the GET_KEY message. On Bob's side, reservation may not have been made, leading to the inability of sending the same GET_KEY request. The concept in ESTI 004 is that the response of GET_KEY is the synchronized key. The application will receive either the synchronized keys or an error message. However, KMS-KMS communication is unavoidable in delivering synchronized keys. Without synchronized communication, there may be a risk of DoS attacks on the KMS with requests to obtain keys, leading to delays in processing requests from other KMS applications. Additionally, the standard implements the QoS timeout specified in OPEN_CONNECT messages. This timeout is used to define the time window available to collect the reserved key. If the timeout expires for a specific key on either side, the key is discarded from the buffers. The timeout should be set to a value which includes the expected time required for successful KMS-KMS communication.

The ETSI 004 standard is exceptional for its reliability in key availability. A range of index values can be specified in GET_KEY messages, which allows the SAE application to obtain a larger number of previously reserved keys without interruption. Although the key reservation process may take longer because of the necessary synchronization of logical buffers in the path (if the buffer has keys which are shorter than the key chunk size values, they must be merged to form a larger key, and if only large keys are available, they must be split to form keys which have the key chunk size value), the application does not have to further consider the availability of keys during operation since this is guaranteed by successful reservation.

We point out that reservation is performed according to the identifiers of the source and the destination SAE application. It thus follows the flow-oriented principle of QoS reservations, where instead of the well-known IntServ 5-tuple identifier (IP addresses of the source and destination, source and destination port numbers and type of transport protocol), unique SAE application identifiers are used. ETSI 004 takes into account the source and destination addresses as Uniform Resource Identifiers (URIs), which include appropriate tokens or identifiers to ensure authentication, authorization and accounting (AAA). The URI can include additional information, such as an OAuth token. The process of the received information can be passed to the subsystem entities in charge of AAA. However, the OPEN_CONNECT message is the only point at which the application is of interest. All further communication within ETSI 004 is based on the KSID value. ETSI 004 does not define the methods of determining unique SAE identifiers, nor does it address their potential collision or misuse.[1]

3.2 Differentiated Services

The Differentiated Service (DiffServ) IP architecture [7–11] was proposed to overcome IntServ issues by introducing scalable and flexible service differentiation between IP flows. A new concept here is considered, which is flow aggregation. At the domain edges, flows result as macro-flows (or traffic aggregates), receiving separate treatment inside the DiffServ domain. In this architecture, each node can select a so-called Per Hop Behavior (more details in Sect. 3.2.2) to operate in accordance with different traffic aggregates (each one characterized by a service differentiation in terms of bandwidth, delay performance, latency and packet loss). DiffServ is a stateless network architecture, and service guarantees cannot be assured. On the other side, it results in a scalable approach to traffic engineering, reducing the burden on network devices and easily scaling as the network grows.

---

[1] In the hypothetical worst case, if a network with a large number of nodes is observed, multiple applications may possess the same SAE because of a lack of precise definitions of how SAE identifiers are determined, leading to a collision.

DiffServ allows users to keep any existing Layer 3 prioritization scheme, including the existing equipment, mixing it with DiffServ-aware nodes. The eight-bit field in the IP header is the DiffServ Code Point (DSCP), which replaces the old definitions of Type of Service (TOS) and Traffic Class (TC), respectively. The DSCP length is one byte, but six bits are enough to manage and select the Per Hop Behavior for each interface. The remaining two bits are reserved for explicit congestion notification and management.

3.2.1   DiffServ Components

In view of the QoS components listed in Sect. 2.3, DiffServ implements the following components:

- Traffic conditioners (policing and shaping): these components define the Traffic Conditioning Agreement (TCA) and perform traffic shaping/policing, guaranteeing that the packets entering the DiffServ domain conform to the TCA and the service provisioning policy of the domain. These components operate at the edges of the DiffServ domain. Shapers instead implement buffering operations (also with packet-loss), increasing the delay of a stream to make it compliant with a particular traffic profile, if needed;
- Packet classifiers: these components are used to assign a specific group to each packet. If the classification has been successfully defined, the packet can be subject to QoS handling in the DiffServ network. Classifiers enable the entire network traffic to be partitioned into multiple classes (with a related priority or service class). Classifiers base their activity on the traffic descriptor of the packet;
- Packet markers: these components allow the classification of a packet based on the DSCP value. Markers add a label into the IP header of the packets by setting the DSCP value to a correct code point. In this way, the packet can be differentiated from others, and a proper QoS group can be assigned to it (the packet is said to be categorized into a particular behavior aggregate). DiffServ also enables alteration of the label, therefore each packet can be re-marked (according to the proper considered policy);
- Schedulers: generally these components are represented by proper traffic queuing algorithms, implemented inside the DiffServ-aware nodes. The most considered scheduling algorithms are Class-Based Weighted Fair Queuing (CBWFQ) [28, 29] and Priority Queuing Weighted Round Robin (PQWRR) [30, 31], which are able to guarantee the right bandwidth level to the different traffic classes;
- Techniques for avoiding congestion: this activity is mainly based on packet dropping algorithms by monitoring network traffic loads to avoid congestion and network bottlenecks. The best known mechanism for avoiding congestion is Weighted Random Early Detection (WRED). Packet droppings are also useful for ensuring compliance of the stream profile to the particular TCA;

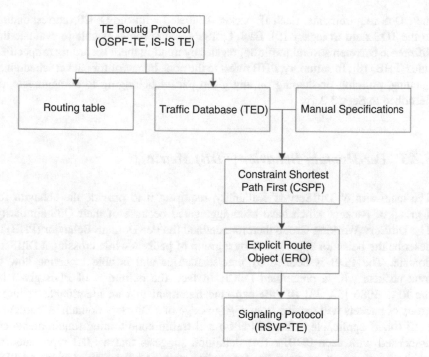

Fig. 3.10 The logical interconnection of DiffServ elements

- Metering activities: the packet classifier sends packets to traffic conditioners which must use a particular metric (or meter) to obtain a measure of the stream's temporal properties to verify whether they comply with the traffic profile from the TCA. Packet markers provide further processing based on whether the packet is in profile or out of profile.

The DiffServ components are interconnected and operate as illustrated in Fig. 3.10

3.2.2   The Per Hop Behavior (PHB) Classes

When any pair of users require a given QoS, it assumes support for end-to-end connection on the inter-domain QoS level. To achieve this goal, many intermediate steps are necessary. The first is support of edge-to-edge QoS between the so-called *ingress* and *egress* of a single network. Therefore, as described earlier, the DiffServ Working Group defined, in the standards, the behaviors required in the forwarding path of all network nodes; they are called Per Hop Behaviors (PHBs). DiffServ is completely based on traffic aggregation: single flows are not routed independently but manipulated and aggregated in a certain number of macroflows according to their QoS requirements. Each IP packet is labeled with the DSCP (corresponding to the TOS field in legacy IP). Each DiffServ-aware router is able to evaluate the difference between several traffic aggregates by treating them according to specified rules (PHB) [8]. In summary, PHB refers to the node behavior for packet scheduling, queuing, policing or shaping on any given packet belonging to an aggregate, as described in Sect. 3.2.1.

3.2.3   Per-Domain Behavior (PDB) Metrics

The main aim of DiffServ is scalability, requiring it to provide the behavior for a group of packets which have been aggregated because of their QoS similarity. The DiffServ Working Group therefore defined the Per-Domain Behavior (PDB) to describe the behavior experienced by a group of packets while crossing a DiffServ domain. The PDB is defined by precise metrics and is able to define how to treat packets with a predefined DSCP. In fact, the definition of PDB given by the RFC 3086 [32, 33] is: "the expected treatment that an identifiable or target group of packets will receive from edge-to-edge of a DiffServ domain. A particular PHB (or, if applicable, a list of PHBs) and traffic conditioning requirements are associated with each PDB.". This definition suggests that a PDB represents the point of interaction between the forwarding path and the control plane (a PDB can be considered the extension of the PHB over an entire DiffServ domain). The parameters (or attributes) of each PDB depend on the characteristics of the traffic aggregate resulting from the classification and conditioning of the packets which enter the DiffServ domain and from the forwarding treatment (i.e., PHB) which the packets receive inside the domain. A very wide range of metrics is always possible for PDB, but generally, they are expressed as bounds or percentiles (instead of absolute values). PDB attributes can be classified into two main categories:

- Long term attributes generally refer to the throughput evaluated over a certain period.
- Short-term attributes generally refer to the permissible burstiness in a traffic aggregate.

The PDB is established and applied to a target group of packets which arrive at the edge of the DiffServ domain. The members of each group of packets are distinguished from other arriving packets by the use of packet classifiers. Through traffic conditioning, an aggregate is created (packets are marked with a DSCP for the PHB), and then the packets undergo the particular treatment related to their *travel* within the DiffServ domain. Five types of PDB are defined in DiffServ:

- Best Effort Per-Domain Behavior (BEPDB): common Internet traffic is sent into a DiffServ network. This involves packets which do not require any special differentiation; they are forwarded only when sufficient resources are available;

- Virtual Wire Per-Domain Behavior (VWPDB): this PDB describes a service identical to dedicated wires between endpoints (i.e., virtual wires). It is especially suitable for any packet belonging to traffic which uses fixed circuits (such as telephones, leased data lines, etc.). This PDB is able to replace the physical wire between two points;
- Assured Rate Per-Domain Behavior (ARPDB): the guarantees for this PDB address only the bit rate and any drop to low probability, without any regard to delay or jitter. With ARPDB, traffic can receive a higher bandwidth than requested, but without any guarantee. There is also a set of specific parameters which must be included with this PDB (Committed Information Rate—CIR, other traffic parameters required for measuring CIR, and the maximum packet size for the aggregate);
- One-to-Any Assured Rate Per-Domain Behavior (OA-ARPDB): this is similar to ARPDB, but it determines only the one-to-any case. This PDB considers the possibility to express the probability that the assured rate will not be met and is useful, for example, when a DiffServ ingress point sends data to any DIffServ egress point;
- Lower Effort Per-Domain Behavior (LEPDB): this PDB is dedicated to non-critical traffic. The basic idea is to delay/drop LEPDB packets if higher priority traffic exists. It is very suitable for multimedia applications or *www* search engines.

3.2.4 ETSI 014: Protocol and Data Format of REST-Based Key Delivery API

Unlike the ETSI 004 standard, which is based on the resource access flow approach, ETSI 014 is based on REpresentational State Transfer (REST) key acquisition communication. REST is an architectural style for distributed organizations which relies on six guiding principles: client-server paradigm, stateless and cacheable communication, uniform interface, layered system, and option code on demand.[2]

In the context of the ETSI 014 standard, use of the REST approach means that there are no dedicated reservations. That is, QoS service is not guaranteed, and the responsibility of providing QoS communication is transferred from KMS entities to SAE applications. Since the communication is performed on the stateless REST principle, it means that there is no session establishment function, use of session identifiers (KSID) nor session closing functions. ETSI 014 defines three API functions:

- GET_STATUS: function to check the status of the available key to the desired SAE destination. The function returns information about the assigned KMS

---

[2] More details about REST can be found at www.restfulapi.net.

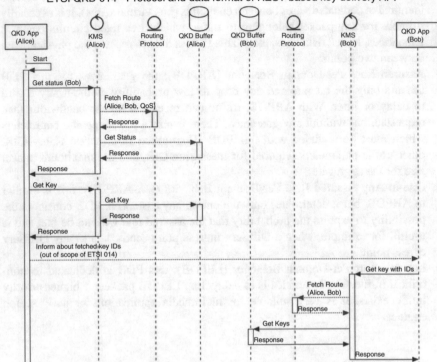

Fig. 3.11 Sequence diagram of ETSI 014—Protocol and data format of REST-based key delivery API. The blue dotted lines is beyond the scope of the ETSI 014 standard [34]

entity for the destination application, the key size that can be delivered to the SAE (bits), the number of keys stored in the key buffers, the maximum number of keys that can be delivered in one response.

- GET_KEY: function for obtaining keys to reach the destination SAE application. The input values conveyed by this message are the number and size of keys requested, additional SAE IDs which can be used to specify two or more slave SAEs sharing the same key, and additional extension parameters. The response is the key container JSON structure, including the key identifier (ID),[3] key encoding using base64, and additional key_extension values.
- GET_KEY_WITH_KEY_IDS: function to obtain keys from KMS for the slave SAE. The input value conveyed by this message is the key identifier ID, while the result should be the same as from the GET_KEY message.

As shown in Fig. 3.11, Alice's SAE application will first request information about the status and possibility of establishing a connection to the destination SAE

---

[3] Note that the keys are organized into blocks, and each block of the cryptographic key is identified using a unique ID. The ID values must be identical on both sides of the QKD link to be managed.

application. After obtaining a response, the application will decide whether and when to request key from the KMS by sending a GET_KEY message. It is up to the application to take into account the obtained status values and calculate the number of keys it will ask for. The maximum number of keys which can be requested from KMS is provided in response to the GET_STATUS message. However, asking for larger quantities of keys per single query can lead to a reduction of further queries and reduce jitter. In that case, the application should store the obtained keys in the local buffer and use them as needed without waiting for subsequent queries and answers from the KMS, which can lead to delays. However, the request to deliver a large number of keys in a single query can lead to increased delay. Due to the enormous size of the content, a large number of IP packets might be fragmented along the path to the destination. In addition, the ETSI 014 approach can lead to a potential collision. The SAE on Alice's side can obtain the key, but the SAE on Bob's side can obtain a response that the key with the defined ID values has already been assigned to another SAE application. To avoid this confusion, clear KMS-KMS and SAE-SAE communication is necessary, however this is beyond the scope of the ETSI 004 and ETSI 014 specifications.

The advantage of the ETSI 014 over ETSI 004 is scalability, as KMS entities do not have to take into account reservation-oriented sessions. With ETSI 004, the key is reserved on the KMSs after successfully processing the OPEN_CONNECT request. However, ETSI 014 does not implement this strict session-oriented key reservation. There is a short period of reservation of the key in the sense that the key which is allocated in response to the GET_KEY request is synchronized and stored for use on both sides. That is, Alice can collect the key, whereas there is no timeout defined on Bob's side to also collect the key. The malign Alice application can thus perform a DoS attack on Bob's KMS. Therefore, ETSI 014 may be more suitable for applications requiring smaller keys or based on periodic key acquisition.

We may notice that neither the ETSI 004 nor ETSI 014 specifications define the priority of the request or the route of the KMSs to establish the key to the destination if the key relaying approach is implemented. The task of defining a route falls under the QKD routing protocol, which is discussed in Chap. 6. However, if ETSI 014 is viewed from the point of view of DiffServ architecture, it can be concluded that one KMS governs the domain of all nodes and devices surrounding it. Therefore, ETSI 014 can attain Per-Domain behavior by contacting all KMSs on the route to the requested destination.

3.3 MultiProtocol Label Switching

Developed by IETF [35] Multi-Protocol Label Switching (MPLS) is an IP technology which is able to overcome the major routing drawbacks of IP. It is used by Internet Service Providers (ISPs) for improved infrastructure dedicated to real-time traffic. MPLS integrates and supports the Frame Relay (FR) and Asynchronous Transfer Mode (ATM) protocols (hence the name multiprotocol), which attempted to introduce several IP enhancements [36]. It deploys FR, ATM and IP to create an Label Switch Path (LSP), increasing also the scalability grade. MPLS has gained great popularity among network operators by enabling simpler management of network flows without modifying routing protocols or changing metrics on specific links. Instead of classic IP routing (packet header analysis, routing table consultation, routing algorithm and forwarding), each packet in MPLS is assigned an forwarding equivalence class (FEC) label. The FEC label should be assigned only once, and it is used to decide on forwarding, completely neglecting the IP lookup operations in each relay node. In the strict sense, FEC denotes a group of packets which are forwarded in the same manner (identical forwarding treatment, over the same or other path).When the packet enters the network, routers use the FEC label as an index to forward the packet with the aid of their own tables [35].

The benefits of MPLS are manifold:

- Multi services over a single infrastructure: MPLS has the capability of integrating several technologies (e.g., Layer2/Layer3 VPNs, QoS, Generalized MPLS, IPv6), with the benefit of enabling scalable and secure networks. As previously mentioned, it is able to combine ATM, FR and IP networks into a single infrastructure, with consequent cost savings for the final clients;
- VPN: implementing a VPN over MPLS is one of the most popular options given by this architecture. It is able to enhance the implementation of private and secure networks (VPNs) over the same network topology, destined to many customers. In this way, it is possible to divide the network into smaller segments, with high scalability features (large enterprises and ISPs are mainly attracted by this feature);
- Scalability: MPLS has been able to completely change classic network topologies generally composed of a large core of ATM switches connected to neighboring routers in a mesh. MPLS ensures that the core nodes are not related to other network and dedicated only to packet forwarding;
- Traffic engineering feature: it refers to the capability of MPLS to optimally control traffic by spreading it across all available links according to different specified metrics.

3.3.1   MPLS Operation and Architecture Basics

The main task of MPLS is to integrate Layer 2 link information (available bandwidth, latency, etc.) into Layer 3 fields, with the main aim of simplifying IP packet forwarding. It accelerates the delivery of packets by assigning packets to the particular FEC. Before a packet arrives in the MPLS domain, Label Edge Routers (LER) classifies the IP packet with an FEC label, as shown in Fig. 3.12. These boundary routers are often denoted Provider Edge Router (PER).

LERs can act both as INGress Label Switching Routers (ING-LSRs) and Egress Label Switching Routers (E-LSRs). ING-LSRs create the MPLS headers and add

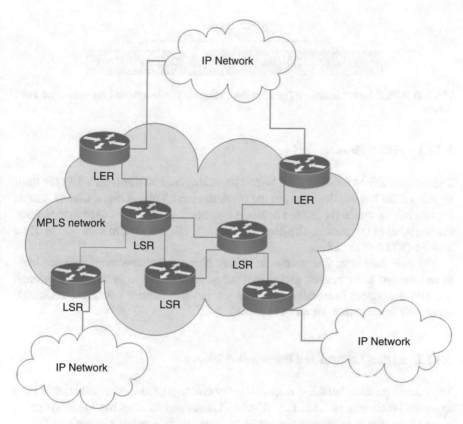

Fig. 3.12 Illustration of the MPLS architecture

them to the IP packets, transferring packets to the data links. At this point, the packets are routed through the LSP by the INTermediate LSR (INT-LSR), also called the Provider Router (PR). LSP is the path comprising the ING-LSR, the INT-LSRs and the E-LSR, therefore it is unidirectional from ING-LSR to the E-LSR. LSR forwarders have the capability of understanding FEC labels and can receive and send labeled packets inside the MPLS network. All the "next-routers" in the MPLS domain will forward the packets without any packet classification operation. Each router uses its own FEC table to identify the packet label. Once it is identified, the label is replaced with the outgoing label and the packet is forwarded to the next INT-LSR. FEC labels have a fixed length, and long prefixes matching the destination IP address are not needed (as for IP). Labeled packets are forwarded up to the final MPLS router (E-LSR), where the labels are removed and the packets are returned as original IP packets and forwarded to the destination address.

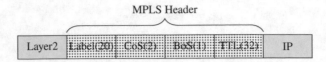

Fig. 3.13 MPLS header structure. The numerical values in brackets denote the size of the fields in bits

3.3.1.1   MPLS Header

The dedicated field for the MPLS header has a length of 32 bits (Fig. 3.13): the most significant 20 bits are dedicated to the label, the next 3 bits define a Cisco Class of Service (CoS), one bit is defined as Bottom of Stack (BoS) and is used to determine the last label in the packet. The least significant 8 bits are used to indicate the Time To Live (TTL) of the label.

The BoS field is very important in MPLS, because it allows more than one label to be assigned to the packet, giving to it the possibility to adequately travel through the MPLS network (generally there is a top label, a bottom label, and a variable number of intermediate labels in between).

3.3.1.2   MPLS Control and Forwarding Planes

The control plane of MPLS is responsible for collecting the information that is used to forward incoming packets. In particular, it sends and receives routing information to and from the neighboring routers. Link-state routing protocols such as OSPF-TE and IS-IS TE are equipped with traffic engineering functionalities to advertise not only the information about the link state (up/down) but also the details of the bandwidth available for LSP. The collected information is stored in the traffic engineering database (TED), which is implemented on each router and used for LSP calculation.

Instead, the MPLS forwarding or data plane is in charge of packet switching after being received at the inbound interface. It is based on the content of labels. Two tables are considered: the Local Information Base (LIB), which contains the map of labels received from neighboring routers, and the Label Forwarding Information Base (LFIB), which is consulted to forward the labeled packets [37, 38]. The LFIB maps the incoming interface and FEC label with the outgoing interface and FEC label. The process of populating the LFIB can be completed through manual configuration or dedicated label distribution protocols. The IETF avoided specifying a single label distribution protocol for MPLS use. However, all protocols allocate labels using the downstream LSR (with respect to the data flow), while the advertisement process is performed by upstream LSR. The FEC values defined in the LFIB on each node define the connection-oriented LSP, which, once it is established, can be used to route traffic along the path. Multiple traffic flows may use the same LSP where the aggregated flow construction is often defined as a traffic trunk.

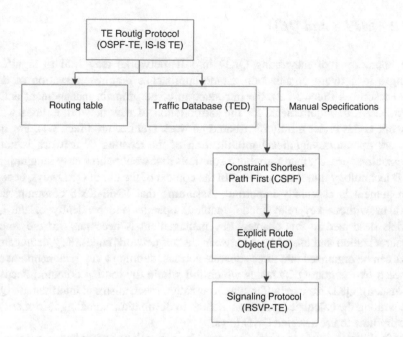

Fig. 3.14 Logical block scheme of MPLS LSP establishment and distribution process

In the LSP definition process, multiple constraints can be applied to path calculation, such as the available bandwidth requested for a particular LSP, inclusion of the administrative attributes of permitted links, the number of hops, LSP setup priority, and others. Generally, constraints can be grouped according to link properties (such as the traffic engineering metric, bandwidth) and LSP properties (such as priority, number of hops, excluded nodes). The collected information from the TED database and specific user configuration values are used by the Constraint Shortest Path First (CSPF) algorithm. As shown in Fig. 3.14 the calculated path is specified in the Explicit Route Object (ERO), which is included in the signaling packet along with the TE-related information. The source node uses a signaling protocol such as RSVP-TE to propagate the ERO specification on each intermediate node along the path. In addition to the list of intermediate nodes which form LSP, the most important information is the metric of interest which intermediate nodes should track and the path setup information, such as LSP priorities. The admission control on each intermediate node will analyze the propagated information and decide whether a successful definition of LSP can be established. If enough resources are available, the path is set up and the available resources on the node are updated [39].

3.3.2 MPLS and QKD

The approaches of integrating QKD into IP networks may lead to significant changes in network organization. Some approaches require reservation of dedicated resources (Sect. 3.1.2), the introduction of per domain management entities (Sect. 3.2.4), or requirements for the introduction of new network entities within network nodes under a fully distributed network architecture (Sect. 3.4). For most network operators, significant modification of the existing IP network is not an acceptable approach, therefore those solutions that would allow easy integration of QKD technology must be identified. In the context of the use of QKD keys, network management is especially important. Assuming that KMS-KMS communication (with or without key relay mode enabled) operates independently of the QoS models described in this chapter, key management is necessary for the smooth synchronization and use of keys between distant network nodes. Key management tasks can be organized into three separate actions: defining a key establishment path between two remote QKD nodes which fall within the routing domain (discussed in Chap. 6); QKD storage/buffer synchronization mechanisms at intermediate QKD nodes along the calculated path from source to destination; signaling information to identify how to use selected QKD keys.

MPLS is a technology widely accepted by network operators because it is a fast and efficient solution for traffic management based on manual requirements and without the need to modify complex routing protocols. MPLS allows the definition of backup paths (color-aware links) which allow additional traffic engineering mechanisms in case of congestion or unexpected events on network segments. Since MPLS is accepted as an efficient management tool, questions arise about how to integrate QKD and MPLS technology.

One approach for the integration of QKD with MPLS/GMPLS was reported in [40, 41]. To facilitate management, the authors proposed the use of a centralized entity which is in charge of defining MPLS LSPs and dedicated cryptographic determinants in ERO headers. The proposed solution gradually directs QKD organization toward a centralized architecture by defining a centralized Path Computation Element (PCE) which collects all statistical and system information from the network. As shown in Fig. 3.15, the source node N1 sends the PCE a request for a quantum encrypted (QE) E2E session for the destination node N5. The authors proposed both active and passive roles for PCE. In active mode, the PCE detects the QE requirements, calculates the optimal path and communicates with the intermediate nodes to configure the path. In passive mode, the PCE will only respond with PCReply to the source node N1 instead of contacting intermediate nodes. Upon receipt, the node will extract the new subjects placed in the ERO which are of value to the QE session (session ID, key length, encryption algorithm, refresh timer settings, and others). Upon analysis of the session ID and conclusion it has no value, the source node will use the ETSI 004 interface to fetch a key from KMS. The source node N1 generates an RSVP Path message for the destination N4, which (assuming no error has occurred) responds with an RSVP Resv message confirming

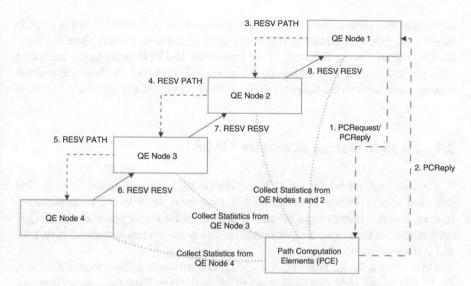

Fig. 3.15 MPLS workflow: establishment of quantum encryption E2E session [40]

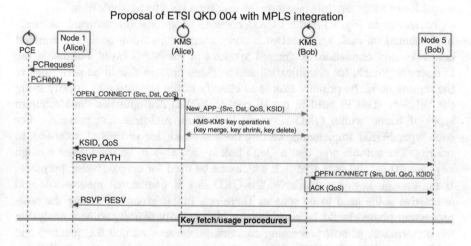

Fig. 3.16 Sequence diagram of combining MPLS RSVP signaling with the ETSI 004 application interface exchanging QoS specifications. The green lines denote KMS-KMS communication, while blue lines denote PCE and RSVP messages

that requested resources have been reserved. The QE service deployment process is complete when the source node N1 receives the RSVP Resv message allowing the application to use keys for cryptographic purposes.

As shown in Fig. 3.16, PCE and KMS entities can be implemented independently, i.e., at different network nodes. MPLS is used to calculate and establish the LSP path between remote distant nodes, while RSVP is to establish the QKD session for exchanging signaling information, such as the session ID, type of cryptographic algorithm, and others. We note that the intermediate nodes which are not QKD aware (N2, N3, N4) do not perform any specific communication since for them the QKD details in the ERO object are irrelevant. The PCE entity in the described example has passive role, although it's active role can lead to direct centralized management, such as the Software Defined Networking (SDN) approach.

3.4   Flexible Quality of Service Model

A Flexible QoS model for QKD Networks (FQKD) was proposed in [27]. The proposed model avoids a centralized management or reservation mechanism. Instead, it uses a distributed approach to control traffic load by providing soft-QoS constraints. No flow or session state information is maintained in support of end-to-end communication.

Driven by the similarities of QKD with ad hoc networks as described in Sect. 2.5, the FQKD model identifies three types of network node: the source node (ingress), the intermediate node (interior) and the destination node (egress). Each node in the network can adopt any role based on the task for a specific network flow.

As shown in Fig. 3.17, the FQKD consists of several fundamental elements implemented on each QKD network node: classifier, waiting queues, admission controller and regulation of granted sessions at the MAC layer. Following the DiffServ approach, the classification and marking process should be performed at the ingress node. Its primary task is to classify traffic according to priority using the DiffServ (DSCP) field in the IP packet. FQKD distinguishes three different types of traffic within QKD network: best-effort, real-time and premium. For each type, FQKD implements waiting queues which are processed according to priority. The authors note that a QKD link is only able to be used when enough keys are exchanged over that link which can be used for cryptographic purposes. If not enough keys are available, the QKD link is considered inaccessible and alternative paths need to be sought. Therefore, traffic generated during the post-processing phase should be considered a high priority. An experiment analyzing the performance of post-processing communication showed that the quantum and public channels of the QKD link are directly connected in a manner that the performance of one channel directly affects the performance of another [27]. Post-processing traffic must therefore not wait for the assignment of keys for cryptographic purposes (which in the post-processing phase are mainly of the authentication type), and this type of traffic must be served as a priority over other types. To more clearly delineate the amount of traffic allocated to a specific priority, the authors proposed implementing thresholds in the token bucket organization as shown in Fig. 3.18. Only traffic from post-processing applications has the right to use key material when $M_{cur,k}(t) \leq M_{min,k}$, while traffic from the other two queues is served only when $M_{cur,k}(t) > M_{min,k}$. The threshold value can be either statically determined or adaptively defined depending on the number of keys in the

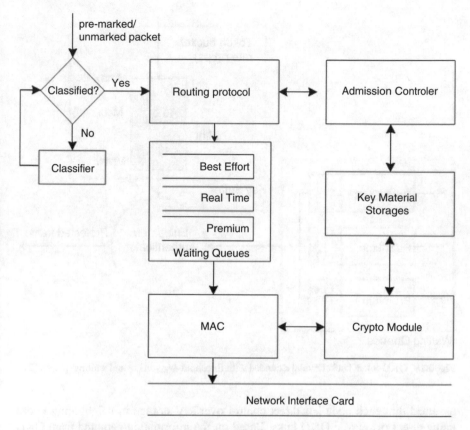

Fig. 3.17 FQKD model organization

QKD stages, the charging/discharging rates and other configuration parameters of the FQKD model.

Because FQKD assumes that no centralized or hierarchical nodes are dedicated or implemented in the network, each node needs to contact a routing protocol to define the route for sending or forwarding packets to the egress node. FQKD is independent of the routing protocol, but in the original version, it relies on the GPSRQ protocol, which we discuss in more detail in Chap. 6. The routing protocol communicates with the admission controller which collects statistics of the available network nodes to decide on the optimal route to the requested destination.

However, according to a typical TCP/IP organization, queues are logically positioned above network interfaces (L2), and FQKD suggests installing queues before contacting routing protocols to more precisely position them on the L4 network layer. The main reason for this is speeding up calculation of network routes and the usability of network links for higher priority traffic.

The FQKD model is the first QoS QKD model for distributed organizations and does not require the implementation of dedicated KMS entities. Instead, it is

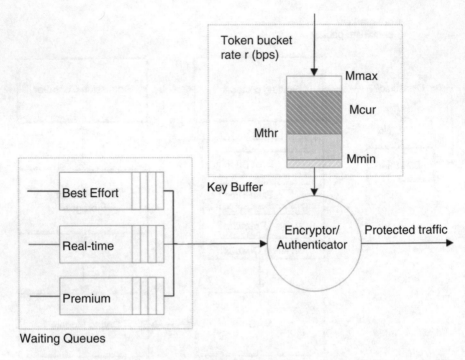

Fig. 3.18 QKD token bucket model extended with threshold M_{thr} value and waiting queues [27]

assumed that each node has direct control over key storage in neighboring nodes using direct or relayed QKD links. Based on the information gathered from QKD post-processing communication, each node can calculate the performance of QKD links, which is vital for defining the optimal routing path.

3.5 Summary

Approaches to satisfying QoS requirements depend on the selected QoS architecture underlying the network's structure. In defining the architecture, the QoS components are selected and their methods of connection and structuring are defined. In this context, both QoS signaling and routing directly relate to the specified QoS architecture because their operation depends on the initially set definitions. From a QoS aspect, it is evident that some of the functionalities can be identified with well-researched QoS models such as IntServ. In the context of QKD technology, which strives for elegant integration with existing IP networks, reservation-oriented and instant server approaches stand out. However, how they are adapted and applied in IP networks largely depends on the volume of traffic sought to be served and the specific characteristics of the network, such as robustness and scalability.

References

1. Mahadevan, I., & Sivalingam, K. M. (1999). Quality of service architectures for wireless networks: Intserv and diffserv models. In *Proceedings Fourth International Symposium on Parallel Architectures, Algorithms, and Networks (I-SPAN'99)* (pp. 420–425). Piscataway: IEEE.
2. Braden, R. T., Clark, D. D., & Shenker, S. (1994). Integrated services in the internet architecture: An overview (1633). https://doi.org/10.17487/RFC1633. https://rfc-editor.org/rfc/rfc1633.txt
3. Clark, D. D., Shenker, S., & Zhang, L. (1992). Supporting real-time applications in an integrated services packet network: Architecture and mechanism. In *Conference Proceedings on Communications Architectures & Protocols* (pp. 14–26).
4. Guerin, R., Partridge, C., & Shenker, S. (1997). Specification of guaranteed quality of service (2212). https://doi.org/10.17487/RFC2212. https://rfc-editor.org/rfc/rfc2212.txt
5. Wroclawski, J. T. (1997). Specification of the controlled-load network element service (2211). https://doi.org/10.17487/RFC2211. https://rfc-editor.org/rfc/rfc2211.txt.
6. Xu, M., Mi, Z., Feng, X., & Xie, W. (2003). Implementation techniques of intserv/diffserv integrated network. In *International Conference on Communication Technology Proceedings, 2003. ICCT 2003* (vol. 1, pp. 231–234). Piscataway: IEEE.
7. Baker, F., Black, D. L., Nichols, K., & Blake, S. L. (1998). Definition of the differentiated services field (DS field) in the IPv4 and IPv6 headers. RFC 2474. https://rfc-editor.org/rfc/rfc2474.txt
8. Black, D. L., Wang, Z., Carlson, M. A., Weiss, W., Davies, E. B., & Blake, S. L. (1998). An architecture for differentiated services. RFC 2475. https://rfc-editor.org/rfc/rfc2475.txt
9. May, M., Bolot, J.-C., Jean-Marie, A., & Diot, C. (1999). Simple performance models of differentiated services schemes for the internet. In *IEEE INFOCOM'99. Conference on Computer Communications. Proceedings. Eighteenth Annual Joint Conference of the IEEE Computer and Communications Societies. The Future is Now (Cat. No. 99CH36320)* (vol. 3, pp. 1385–1394). Piscataway: IEEE.
10. Jacobson, V., Nichols, K., & Poduri, K. (1999). An expedited forwarding PHB. Request for Comments (Proposed Standard) 2598.
11. Weiss, W., Heinanen, J., Baker, F., & Wroclawski, J. T. (1999). Assured forwarding PHB group. RFC 2597. https://rfc-editor.org/rfc/rfc2597.txt
12. Clark, D. D., & Fang, W. (1998). Explicit allocation of best-effort packet delivery service. *IEEE/ACM Transactions on networking, 6*(4), 362–373.
13. Floyd, S., & Jacobson, V. (1993). Random early detection gateways for congestion avoidance. *IEEE/ACM Transactions on Networking, 1*(4), 397–413.
14. Guerin, R., & Peris, V. (1999). Quality-of-service in packet networks: Basic mechanisms and directions. *Computer Networks, 31*(3), 169–189.
15. Stoica, I., Shenker, S., & Zhang, H. (2003). Core-stateless fair queueing: A scalable architecture to approximate fair bandwidth allocations in high-speed networks. *IEEE/ACM Transactions on Networking, 11*(1), 33–46.
16. Chen, R., Yilmaz, O., & Yen, I.-L. (2006). Admission control algorithms for revenue optimization with QoS guarantees in mobile wireless networks. *Wireless Personal Communications, 38*(3), 357–376.
17. Choi, S., & Shin, K. G. (2002). Adaptive bandwidth reservation and admission control in QoS-sensitive cellular networks. *IEEE Transactions on Parallel and Distributed Systems, 13*(9), 882–897.
18. Zhao, D., Shen, X., & Mark, J. W. (2002). QoS performance bounds and efficient connection admission control for heterogeneous services in wireless cellular networks. *Wireless Networks, 8*(1), 85–95.
19. Zhu, C., Pei, C., Li, J., & Kou, W. (2005). QoS-oriented hybrid admission control in IEEE 802.11 WLAN. In *19th International Conference on Advanced Information Networking and Applications (AINA'05) Volume 1 (AINA Papers)* (vol. 1, pp. 484–487). Piscataway: IEEE.

20. Boyle, J., Cohen, R., Durham, D., Herzog, S., Rajan, R., & Sastry, A. (2000). The cops (common open policy service) protocol. Request for Comments (Proposed Standard), 2748.
21. Karlsson, M., Karamanolis, C., & Chase, J. (2005). Controllable fair queuing for meeting performance goals. *Performance Evaluation, 62*(1–4), 278–294.
22. Parekh, A. K., & Gallager, R. G. (1993). A generalized processor sharing approach to flow control in integrated services networks: The single-node case. *IEEE/ACM Transactions on Networking, 1*(3), 344–357.
23. Dovrolis, C., Stiliadis, D., & Ramanathan, P. (2002). Proportional differentiated services: Delay differentiation and packet scheduling. *IEEE/ACM Transactions on Networking, 10*, (1), 12–26.
24. Braden, R. T., Zhang, L., Berson, S., Herzog, S., & Jamin, S. (1997). Resource ReSerVation protocol (RSVP) – Version 1 functional specification. RFC 2205. https://rfc-editor.org/rfc/rfc2205.txt
25. Wroclawski, J. T. (1997b). The use of RSVP with IETF integrated services. RFC 2210. https://rfc-editor.org/rfc/rfc2210.txt
26. ETSI ISG QKD. (2020). Quantum key distribution (QKD); application interface. 2, 1–22. https://www.etsi.org/deliver/etsi_gs/QKD/001_099/004/02.01.01_60\penalty-\@M/gs_qkd004v020101p.pdf
27. Mehic, M., Fazio, P., Rass, S., Maurhart, O., Peev, M., Poppe, A., Rozhon, J., Niemiec, M., & Voznak, M. (2020). A novel approach to quality-of-service provisioning in trusted relay quantum key distribution networks. *IEEE/ACM Transactions on Networking, 28*(1), 168–181. ISSN 1063-6692. https://doi.org/10.1109/TNET.2019.2956079
28. Badr, S., Bayoumi, F., & Darwesh, G. (2011). QoS adaptation in real time systems based on CBWFQ. In *2011 28th National Radio Science Conference (NRSC)* (pp. 1–8). Piscataway: IEEE. https://doi.org/10.1109/NRSC.2011.5873626
29. Kwiatkowski, M., & Elliot, M. (2003). Using router diffserv mechanisms to implement military QoS. In *IEEE Military Communications Conference, 2003. MILCOM 2003* (vol. 2, pp. 960–965). Piscataway: IEEE.
30. Mao, J., Moh, W. M., & Wei, B. (2001). PQWRR scheduling algorithm in supporting of diffserv. In *ICC 2001. IEEE International Conference on Communications. Conference Record (Cat. No. 01CH37240)* (vol. 3, pp. 679–684). Piscataway: IEEE. https://doi.org/10.1109/ICC.2001.937326
31. Yang, M., Lu, E., & Zheng, S. Q. (2003). Scheduling with dynamic bandwidth allocation for diffserv classes. In *Proceedings. 12th International Conference on Computer Communications and Networks (IEEE Cat. No. 03EX712)* (pp. 319–324). Piscataway: IEEE.
32. Nichols, K., & Carpenter, B. E. (2001). Definition of differentiated services per domain behaviors and rules for their specification (3086). https://doi.org/10.17487/RFC3086. https://rfc-editor.org/rfc/rfc3086.txt.
33. Nichols, K., & Carpenter, B. E. (2001). Definition of differentiated services per domain behaviors and rules for their specification. RFC 3086. https://rfc-editor.org/rfc/rfc3086.txt
34. ETSI ISG QKD. (2019). Quantum key distribution (QKD); protocol and data format of REST-based key delivery API (ETSI GS QKD 014 V1.1.1) 1, 1–22. https://www.etsi.org/deliver/etsi_gs/QKD/001_099/014/01.01.01_60/gs_qkd014v010101p.pdf.
35. Viswanathan, A., Rosen, E. C., & Callon, R. (2001). Multiprotocol label switching architecture. RFC 3031. https://rfc-editor.org/rfc/rfc3031.txt
36. Duangkreu, W., Kerddit, S., & Noppanakeepong, S. (2000). Frame relay to atm PVC network interworking management. In *2000 TENCON Proceedings. Intelligent Systems and Technologies for the New Millennium (Cat. No. 00CH37119)* (vol. 1, pp. 522–526). Piscataway: IEEE.
37. Lopez, G., & Grampin, E. (2017). Scalability testing of legacy MPLS-based virtual private networks. In *2017 IEEE URUCON* (pp. 1–4). Piscataway: IEEE.
38. Das, S., Sharafat, A. R., Parulkar, G., & McKeown, N. (2011). MPLS with a simple open control plane. In *Optical Fiber Communication Conference* (p. OWP2). Washington: Optical Society of America. https://doi.org/10.1364/OFC.2011.OWP2

39. Minei, I., & Lucek, J. (2010). *MPLS-enabled applications: Emerging developments and new technologies*. Hoboken: Wiley.
40. Aguado, A., Lopez, V., Martinez-Mateo, J., Peev, M., Lopez, D., Martin, V., Lopez, V., Martinez-Mateo, J., Peev, M., Lopez, D., & Martin, V. (2017). GMPLS network control plane enabling quantum encryption in end-to-end services. In *2017 International Conference on Optical Network Design and Modeling (ONDM), Budapest, Hungary*, 645127, pp. 1–6. Piscataway: IEEE. ISBN 978-3-901882-93-7. https://doi.org/10.23919/ONDM.2017.7958519
41. Aguado, A., Lopez, V., Lopez, D., & Martin, V. (2017). Experimental validation of an end-to-end QKD encryption service in MPLS environments. In *Qcrypt 2017, Cambridge* (pp. 3–5).

Chapter 4
Quality of Service Media Access Control of Quantum Key Distribution Networks

The conceptual composition of a QKD link into quantum and physical channels corresponds to similar connections in the physical world and does not preclude the dual use of fibres as both a classic and quantum channel (in fact, the SECOQC network was built on existing fibre-optic cables). Viewed from a logical perspective, the network consists of more layers than just these two channels. At the lowest layer is a physical optical infrastructure which dictates the connectivity of quantum channels. The second layer contains an IP network through which post-processing point-to-point operations are performed. The third logical layer includes key relay and key management operations for the purpose of delivering keys to remote network locations. It is implemented through the same public channels over which post-processing traffic is transmitted (some of the combinations of these two layers are discussed in Chap. 5), but logically it is a separate layer. The top layer contains applications which use keys to protect traffic exchange. Considering the progress which has been achieved in combining quantum and public channels over the same fibre [1–3], practical communication can be performed over the same network infrastructure. However, logical organization of the network remains as a result of the specificity of each of these layers.

The process of key production necessitates its separation from the process of key use. From the point of view of applications which consume keys, all lower layers may be considered infrastructure layers. Specifically, these applications require cryptographic keys and do not dive very deeply into the mechanisms of establishing keys or the problems which occur at lower layers. Following from the similarity between QKD and MANET networks described in Sect. 2.5, the first three logical layers in QKD networks can be similarly viewed in terms of the Media Access Layer (MAC) in WiFi networks (Fig. 4.1).

In this book, we do not deal with the phenomena occurring at the optical/quantum layer; instead, we focus on the network aspect involving the upper layers. We refer interested readers to references [4–8] for detailed discussions of quantum layer performance.

© Springer Nature Switzerland AG 2022
M. Mehic et al., *Quantum Key Distribution Networks*,
https://doi.org/10.1007/978-3-031-06608-5_4

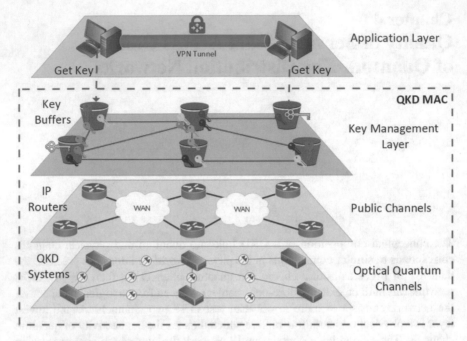

Fig. 4.1 Logical organization of a QKD network

4.1 Post-Processing Applications

After a raw key is generated by exchanging qubits over a quantum channel, all further communication, known as post-processing, is performed over a public channel. Although differences exist in implementation, almost every QKD protocol includes the following post-processing steps: extraction of the raw key (sifting), quantum bit error rate (QBER) estimation, key reconciliation, privacy amplification and authentication [6].

Different key generation processes such as QKD protocol and post-processing techniques can result in different volumes of network traffic. If a dedicated connection is not applied, post-processing traffic-flows may be affected with the congestion caused by other network applications.

Let us suppose that QKD components are connected sequentially, as shown in Fig. 4.2. Because the QKD key comes in blocks, the QKD optical components must wait until post-processing has completed (post-processing of the next block of keys cannot be performed until the previous block has been processed). The duration of post-processing may depend on route congestion or other issues in the IP network (if no dedicated line for post-processing is applied), and therefore the performance of the public channel may directly affect the quantum channel. In practice, this type of limitation can be avoided by parallelizing QKD systems, pipelining the steps or using multiple hardware machines to perform post-processing.

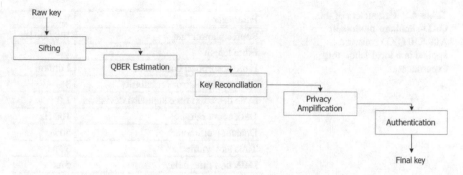

Fig. 4.2 Sequential organization of QKD post-processing modules

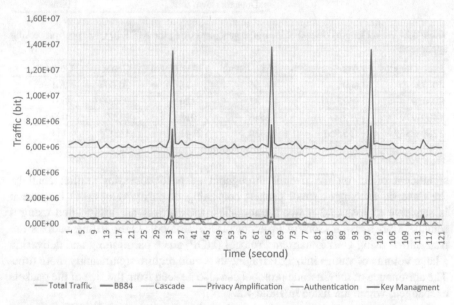

Fig. 4.3 Post-processing traffic "heart-rate" pattern generated using the AIT R10 post-processing tool [9]

Figure 4.3 shows the amount of traffic generated during post-processing using the BB84 protocol and the AIT R10 QKD post-processing tool [9]. Periodic repetition of the "heart-rate" pattern begins with the exchange of sifting information and error-reconciliation processing which dominates the overall amount of network traffic. In 2016, an experiment to measure traffic on the public channel established between two servers on a local network was performed using an AIT QKD-simulator module to continuously generate raw key material of 3% QBER which was then stored on a local hard-drive in advance. The parameters of the QKD-simulator software are listed in Table 4.1.

Average throughput was 6.57 Mbps, with periodic peaks of 13.9 Mbps repeated every 30 seconds as result of sifting. Figure 4.3 shows that the Cascade has a

Table 4.1 Parameters of the QKD-simulator module in AIT R10 QKD software applied in a local laboratory experiment

Parameter	Value
Source photon rate	100,000 Hz
Fibre length	1 km
Fibre absorption coefficient	1 db/km
Source signal error probability	3%
Sync detection time standard deviation	1.0
Dark count rate	100 Hz
Detection efficiency	50%
Time slot width	30 ns
Detection time delay	5 ns
Detection time standard deviation	1
Detector down time	10 ns

Table 4.2 Packet lengths (bytes) of the traffic generated using the AIT R10 QKD post-processing application

Packet lengths	Average length	Min length	Max length	Count	Count (%)
40–79	6604	66	78	20,067	201
80–159	9623	82	126	966,078	9661
640–1279	79,643	646	1018	2728	027
1280–2559	155,952	1320	2141	11,076	111

significant effect on the overall throughput of post-processing traffic. Namely, the Cascade exchanges small packets consisting of parity values, but because it is sensitive to loss and network anomalies, the packets are transmitted using a reliable Transmission Control Protocol (TCP) protocol. Only sifting packets are transmitted using User Datagram Protocol (UDP) since partitioning and delivering a large volume of values into TCP fragments would require significantly more time. The dominance of the Cascade protocol can also be seen from the size of the packets exchanged, which are listed in Table 4.2.

To analyze the performance in a real environment, a virtual QKD link over publicly available IP infrastructure was conducted. Since IP traffic can take arbitrary routes, variations in link delay and bandwidth were noted over one month of testing. Figure 4.4 shows the overall post-processing throughput, with several noticeable variations. The throughput in this real networkhows the amount of traffic gen was significantly lower than the throughput indicated in Fig. 4.3, specifically, the delays and congestion in post-processing traffic were reflected in the quantum channel. Since the modules are connected in series, sifting operations were not able to commence until the previous key block had been processed.

To examine the effects of the quantum channel on the public channel, the authors analyzed the throughput of post-processing for different QBER values. The experiment also included the link bandwidth reduction tool *wondershaper*, which provides analysis of the post-processing application's performance with a smaller permissible bandwidth [10]. The tool reduced the link's bandwidth every

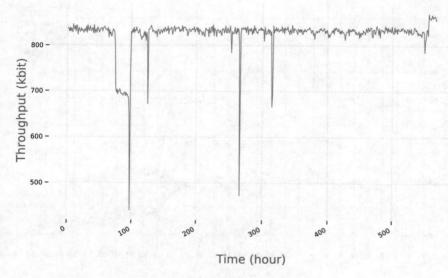

Fig. 4.4 Throughput of the QKD post-processing application over the link between AIT Vienna and VSB-TUO Ostrava for a period of 568 hours, recorded in September 2016

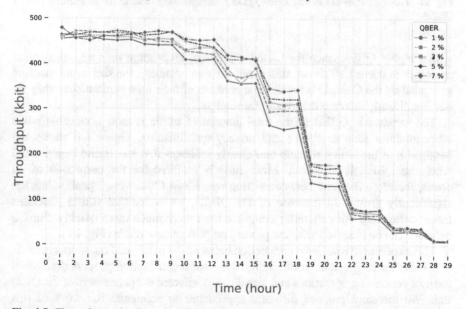

Fig. 4.5 Throughput of a Cascade QKD post-processing module for different values of QBER. The link's bandwidth was reduced every three hours to analyze the robustness and resilience of the post-processing application

three hours. Figure 4.5 shows the bandwidth of the Cascade module for different QBER values. Although it is intuitive to expect a larger volume of traffic for cases

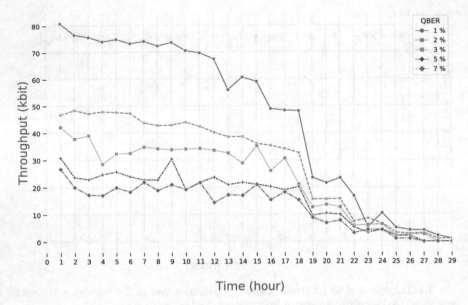

Fig. 4.6 Throughput of the sifting BB84 QKD post-processing module for different values of QBER

with a higher QBER since the Cascade protocol takes more time and exchanged messages to correct all errors, this is not the case. Namely, the size of the packets generated by the Cascade is small, so regardless of their increased number, they do not significantly increase the overall throughput.

The increase in QBER to the total amount of traffic is most noticeable when other modules such as sifting and privacy amplification. Figure 4.6 shows the bandwidth of the sifting module and clearly indicates that less traffic is generated when the QBER is increased. More time is required for the correction of all errors, therefore the time between sifting operations ("heart-rate" peak values) is significantly greater. In the case of 1% QBER, we notice that sifting generates more traffic since other modules complete their work much more quickly. Similar behavior can be observed with the privacy amplification module (Fig. 4.7).

This experiment revealed a connection between the public and quantum channels which cannot be ignored. Additionally, the error reconciliation process dominated the post-processing operations and significantly affected the parameters of the QKD link. We therefore propose different approaches to accelerate the work of this module.

The simplest approach in accelerating the performance of post-processing operations is by operating the post-processing modules in parallel. Maurhart proposed several separate post-processing modules which are able to process the work in parallel [9]. While one Cascade module processes the block of one key, another Cascade module can process another block. Organization in this manner would not slow down the operation of optical quantum devices, which under serial connection

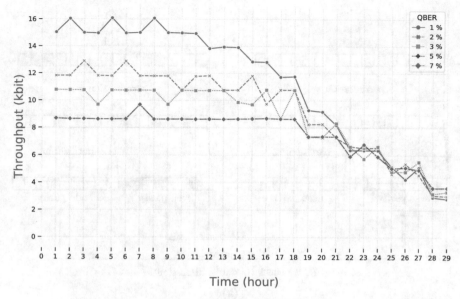

Fig. 4.7 Throughput of the privacy amplification QKD post-processing module for different values of QBER

must wait for release of the memory occupied by processing the previously delivered qubits. In the following sections, we consider approaches that, in addition to increasing the security of the MAC layer, also affect its performance.

4.1.1 Improving Error Reconciliation

Since the eavesdropping detection mechanism of QKD relies on a reliable estimate of errors, whose unexpected increase would reveal the presence of an adversary, we cannot easily apply standard methods such as error correction codes or similar. The Cascade protocol as described in Sect. 1.2.1.3 was designed to estimate and correct a potentially large number of errors, but does so based on fixed block partitioning. Let us now briefly reconsider the block length choice.

Figure 4.8 shows a simplified view of several stages, and respectively, repetitions, of the Cascade protocol, along which the presence of errors is detected through parity mismatches, with a recursive narrowing of the error's region by repeating the protocol on smaller and smaller portions of the key stream (Fig. 4.8b).

Quite evidently, parity errors are only detectable if they appear in odd counts, and we can ask for the likelihood of missing an even number of errors by observing parities only. The purpose of the permutation π shown in Fig. 4.8a is precisely to keep the chances of missing an error low, but if we can reasonably impose an error scattering model, we can also attain some analytic results.

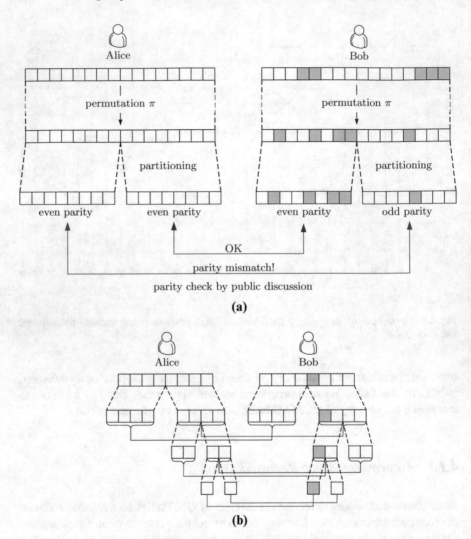

Fig. 4.8 Cascade protocol (abstract simplified view). (**a**) Single round of cascade to determine block parity (mirrored view between Alice and Bob). (**b**) Recursive repetition to correct the error

Specifically, we may think of a "natural" error event in the bit stream as following a Poisson process, with the rate parameter being precisely the expected QBER. Based on measurements taken from lab prototypes [11], this rate was empirically found to be roughly constant yet generally affected by external factors (excluding the adversary).

If the QBER undergoes natural variations which are known from measurements on the "unattacked" line, we can derive a model for error scattering according to a *Cox process*. That is, if the error rate is constant, then we expect a number $N \sim Pois(QBER)$ of errors per unit of time. The Cox process merely replaces the constant error rate $QBER$ with one that varies over time. This variation can be a deterministic, perhaps even periodic, function, or change randomly.

To illustrate, suppose that the QBER changes with time over the day, for example, because of vibrations of traffic flowing over the streets where the cables are buried, or it changes because of weather conditions or pollution levels which rise during traffic peak times (in the case of wireless QKD links). The QBER rate parameter may then depend on time, and we have an inhomogeneous Poisson process, i.e., a Cox process, which inserts a random number of N errors per block.

4.1.1.1 Adaptive Cascade

Equipped with an error scattering model, let us now return to Cascade and ask how likely it is to have an even number of errors in a block that would be lost through a parity check. Some concrete numbers can be obtained for specific error scattering models, for example, if the error rate is Gamma-distributed as $QBER \sim Gamma(a, b)$ with parameters $a, b > 0$, then the chance for an odd number of errors is [12]:

$$\Pr(\text{odd number of errors}) = \frac{1}{2}\left(1 - \left(\frac{b}{b+a}\right)^a\right),$$

(with a related yet more complicated formula if the string is assumed to have only finite length). The initial block size based on this particular Cox process instance comes to

$$\text{initial block size} \approx \frac{f}{a/b},$$

where f is the overall frequency of qubits per unit of time. Generally, the recommendation here is the following rule for selecting the initial block size, according to the steps:

1. given an error scattering model which lets us expect (on average) a number of N errors,
2. set the initial block length inverse-proportionally to the number of errors per unit of time so that at most one error per block occurs or is expected per unit of time. Figure 4.9 graphically illustrates the idea.

The error scattering model described above is only an option and may or may not practically be valid under the specific environmental conditions of a particular QKD link. The general recommendation is therefore to craft the error scattering model according to the specific QKD link environment, if possible. Allowing the block size to vary according to an error scattering model is known as *adaptive Cascade* [13].

bit error density

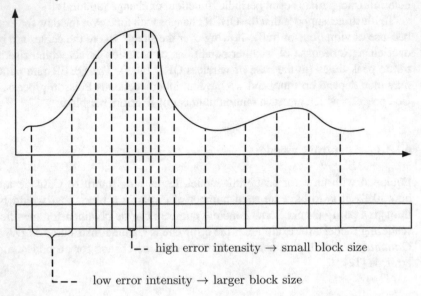

high error intensity → small block size

low error intensity → larger block size

Fig. 4.9 Adapting the block size to the expected error counts

4.1.1.2   Using Error Correcting Codes

If the error scattering model permits a reliable estimate, one could think further by making error correction on both sides completely non-interactive by:

- partitioning the incoming stream into blocks which contain at most (on average) one bit error,
- treating the blocks as words in the error correction code so that the same corrected string can be computed at both ends of the channel without any direct interaction (cf. the Winnow protocol).

Following this intuitive line of thought and continuing the previous error scattering model as a Cox process with a Gamma-distributed rate parameter, suppose that we perform error correction through a conventional block code of Hamming distance d. That is, we adapt the block size not to contain one error on average but with a high likelihood of containing up to the maximum number of errors that the block code can correct (see Fig. 4.10)

If the random variable X is the actual error count in the block (following the Cox process), a linear block code can generally repair a fraction of $N \cdot \Pr(X \leq \lfloor (d-1)/2 \rfloor)$ bits, leaving a residue of

$$N \cdot \Pr(X > d - 1) \tag{4.1}$$

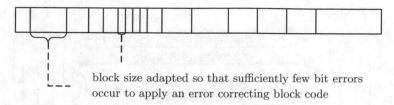

block size adapted so that sufficiently few bit errors
occur to apply an error correcting block code

Fig. 4.10 Adaptation of the block size to a maximum error count

bits to be interactively "repaired" by the Cascade protocol. To perform correction
non-interactively, we require a bound for the remainder probability in (4.1).
Specifically, given the distribution of X from the error scattering model (Poisson
or Cox process) and selecting some small value $\varepsilon > 0$, we can look for the smallest
integer t such that $\Pr(X > t) < \varepsilon$. Given this value t, we can design a block code
of Hamming distance $d = 2t + 1$ to correct up to t expected errors, with a chance
of $< \varepsilon$ in missing an error. We can then non-interactively correct up to $(1 - \varepsilon) \cdot N$
out of N bits. Referring to [14] for details of the calculation, an upper bound for t,
based on a $Gamma(a, b)$-distributed QBER is given by

$$t \leq \max\left\{1, \ln\left[\frac{2b^a}{\varepsilon}\right] - a\ln\left[b + N - \sqrt{N(b+N)}\right]\right\}.$$

It is fair to remark that (at the time of writing this book) we are unaware of
any practical exploration or exploitation of this possibility apart from the Winnow
protocol, which uses Hamming codes. In this context, it is also important to bear in
mind that all this assumes the absence of an adversary from the protocol, i.e., error
correction relies on the accuracy of the error scattering model. An adversary may
most likely invalidate this error scattering pattern, thereby forcing the correction
process to deliver incorrect results. However, the resulting effect is essentially the
same as the adversarial indicator, because it simply produces an unusually high error
count in the final string, upon whose recognition Alice and Bob would terminate the
protocol. The difference here is only that the error count is no longer produced by
Cascade error correction but must be determined through other means. One such
possibility is out-of-band authentication by comparing parts of the exchanged key.
If the adversary attempts to listen or a person is in the middle, the bit strings which
Alice and Bob compare should still exhibit an overly large number of errors. We
describe this overall simple idea in a few possible examples in the next section.

4.1.2   Out-of-Band Authentication and Key Validation

As in other contexts, we can use two-factor authentication in QKD to improve
the reliability of checks. In this particular case, Alice and Bob could connect on

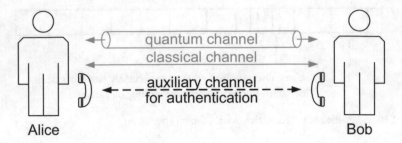

Fig. 4.11 Out of band key validation between alice and bob

a separate channel which is, by assumption, not intercepted by Eve, and discuss how their keys would look to detect whether an adversary is listening, or worse, whether there a person is in the middle. This concept, visualized in Fig. 4.11 has two assumptions:

1. the existence of a third channel which Eve cannot intercept, in addition to the public negotiation and quantum channels,
2. the protocol does not allow either Alice or Bob to have full control over the resulting key.

Both assumptions are nontrivial and need careful consideration in practice. The second requirement is known as *key agreement* in the related literature, where both parties have equal influence over the resulting key (a classic example is the Diffie-Hellman protocol, assuming an implementation which is not vulnerable to small-subgroup attacks). The complementary concept of *key exchange* would let either Alice or Bob determine the key which the other respective party uses.

To reliably detect a person in the middle, we need to assume that Eve, interacting with both Alice and Bob, does not have full control over the keys she establishes with both. Although most likely being so in many instances of QKD, for this assumption to be provably reliable requires explicit measures to this end. We describe two such possibilities in the following section, both which can be characterized as *watermarking* the qubit stream in a QKD protocol.

4.1.2.1   Watermarking by Quantum Coin Flipping

The main problem of coin flipping (over the phone) is an interaction between two parties, Alice and Bob, in which Bob tosses a coin and asks Alice to guess. Obviously, without Alice seeing Bob's action, and hence unable to verify whether what Bob does is honest, Bob can always call Alice's guess incorrect. A coin flipping protocol is used to prevent Bob from cheating and to enforce fairness. Its application to QKD has been described in [15], proposing that Alice and Bob could interleave an instance of BB84 with an instance of the quantum coin flipping protocol detailed in [16]. This combination is especially elegant since the coin

1. Alice sends a random sequence of BB84 states to Bob.
2. Bob measures the received qubit on a randomly selected basis.
3. Bob tells Alice which basis he used for each photon he received.
4. Alice tells Bob which of these bases were correct.
5. Alice and Bob keep only the data from these correctly measured photons, discarding the remainder.
6. The data is interpreted as a binary sequence according to the coding scheme. Error correction and privacy amplification form the final key.

(a)

1. Alice transmits a randomly selected BB84 state $|\psi_{b,x}\rangle$ to Bob, where b is the basis and x is the bit.
2. Bob measures the received qubit on the randomly selected basis $\hat{b}$. If his apparatus does not register anything, he requests Alice to send another qubit, otherwise let $\hat{x}$ denote the measurement outcome.
3. Bob sends a randomly selected classic bit g to Alice.
4. Alice reveals her original b and x to Bob.
5. If $b = \hat{b}$ and $x \neq \hat{x}$, then Bob aborts the protocol, declaring Alice a cheater; if $b \neq \hat{b}$, then he has no way to verify Alice's honesty.
6. If Bob did not abort the protocol, then the outcome of the coin flip is $x \oplus g$.

(b)

Fig. 4.12 Similarities between BB84 and a quantum coin flipping protocol. (a) BB84 quantum key distribution [17]. (b) Quantum coin flipping [16]

flipping protocol *almost* looks like BB84; in fact, it is only a slight variation of the protocol toward achieving an entirely different goal, namely to establish a bit shared by Alice and Bob but with the assurance that neither has ultimate control over the outcome of the bit. This would also apply to Eve as a person in the middle, thus rendering her detectable if Alice and Bob, out of band, compare their keys. Figure 4.12 shows both protocols beside each other to display the striking similarity.

Interleaving of the two protocols is established by randomly switching between BB84 and coin flipping, based on a pseudorandom sequence that Alice secretly maintains during her conversation with Bob. Only once the protocol has finished can she reveal her seed of the sequence to Bob to allow him to reconstruct the positions where Alice was running the BB84 or coin flipping. Once the coin-flipped bits are known to Bob, these (and only these) would be compared to Alice's outcomes on her side to detect the eavesdropper, who, owing to coin flipping, could not have enforced equality between Alice and Bob.

The protocol for Alice to authenticate herself to Bob is then as follows:

Initialization Alice and Bob agree on a security parameter $\delta \in (0, 1)$ and an unconditionally secure coin-flipping scheme with bias $0 < \varepsilon < \frac{1}{2}$. They select a prime p such that $2^{n-1} < p < 2^n$, where $n > \frac{\log \delta - \log 3}{\log(1/2+\varepsilon)}$, and securely pre-distribute random numbers $r_1, r_2 \in \mathbb{Z}_p$. A pre-shared secret random seed s is assumed available for a pseudo-random bit generator.

Protocol (Guaranteeing a False-Positive Authentication Probability $< \delta$)

1. Alice and Bob use the BB84 protocol to establish a key with their respective partner (also possibly Eve) but interleave the protocol with the coin-flipping protocol given by [16]. The decision when to switch is made through the pseudorandom bit-sequence arising from the seed s.
2. Alice uses the outcome of their repeated coin-flips to randomize the keys shared with their respective counterparts.
3. Alice uses n bits from her (now further randomized) QKD key, calls the bit-string a and calculates $h_1 = r_1 a + r_2$ over $\mathbb{Z}_p$, where the bits of a are treated as a binary representation of a number in $\mathbb{Z}_p$. She sends the result h_1 to Bob.
4. Bob verifies whether he obtains the same result locally and accepts Alice's identity upon equality.

Clearly, this method strongly hinges on the similarity between the QKD and coin flipping protocols, and hence only applies to these specific methods. The overall scheme, however, can be generalized in a straightforward manner.

4.1.2.2 Watermarking Using Pseudorandom Sequences

Dispensing with the coin flipping concept and considering the third channel which Eve does not control, Alice and Bob could simply embed pseudorandom bits at pseudorandom positions in their qubit stream [18]. That is, Alice uses two pseudorandom sequences, one to determine whether she polarizes her qubit to encode 0 or 1, the other to determine whether a real random qubit or a pseudorandom qubit has been inserted into her stream with Bob. For authentication, Alice later tells Bob over an auxiliary channel what her seeds were for the pseudorandom sequence so that Bob can check whether he received the right measurements according to the information Alice sent at the respective positions. This requires significant book-keeping on which bits have been lost and re-sent, complicating the implementation slightly. It also implicitly assumes that Eve is incapable of distinguishing pseudorandom from truly random bits in a sequence. If Alice had infinite computational power, she could check all exponentially many subsets of the stream and recognize the pseudorandom bits among the truly random bits which Alice transmits. Realistically though, this is unlikely to happen unless Eve manages to predict or hijack Alice's random generator. But assuming this possibility is close to assuming that Eve has direct access to Alice's devices, in which case the overall protocol is broken from the beginning.

We refrain from repeating a full description of the process here because of its similarity to the authentication protocol from the previous section. The only difference is the use of two pseudorandom sequences to determine the positions and values of the interleaved qubits. The more important advantage is that the method is sufficiently general to be integrated into more QKD protocols than only BB84, as the change is only to either use truly random or pseudorandom bits or to perform more thorough bookkeeping to later recognize the bits which should have been pseudorandom in the final output if they come from Alice. The other aspect is the implicit introduction of computational intractability due to pseudorandom bits being indistinguishable from truly random bits. The "pseudo" is important in allowing Bob to reconstruct the watermark at the other end. Of course, the idea suggests its own generalization by using more advanced robust watermarking schemes to authenticate the qubit stream. We leave this idea with the pointer in that direction and do not pursue deeper discussion.

The other assumption of using more than one public channel can also be generalized to two or more channels connecting Alice and Bob. This scenario can be used to increase security or even establish end-to-end security where QKD initially only provides point-to-point key exchange. Chapter 7 explores this concept in more detail.

4.2   Overlay QKD Networking

To avoid the effect of public channel congestion on the performance of post-processing operations, different approaches can be considered. One of these approaches involves implementing an overlay network through which post-processing communication is accomplished.

Network domains (autonomous systems) typically communicate routing information using the Border Gateway Protocol (BGP), which is known to be slow in reacting to unexpected network events [19]. Measurement experiments have indicated that following a failure, BGP can take tens of minutes to obtain a coherent representation of the network topology [20]. This can have significant consequences for real-time traffic which cannot tolerate waiting for alternative routes to pass, and since BGP advertises only one route, network nodes are precluded from viewing alternate pathways, including those they may prefer.

An overlay network uses underlying network services to enable implementation of improved services, and the independence of the path given by an Internet Service Provider (ISP) is one of its most significant characteristics (Fig. 4.13). The main aim is to identify alternate routes and avoid low quality paths, switch communications rapidly over multiple routes or even use many routes simultaneously (the application of multipath QKD communications is discussed in Chap. 7).

An ISP measurement study has shown that almost 90% of point-of-presence pairings have at least four link-disjoint routes connecting them [21, 22]. The overlay network therefore allows measurement of the performance of those paths (latency,

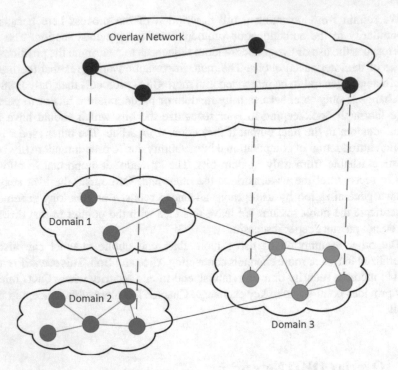

Fig. 4.13 Illustration of an overlay network

availability, throughput and other parameters) to determine the optimal route or perform load balancing at different times [23–26].

Overlay networks are also popular for adding new functionalities which are difficult to implement in conventional IP networks. For example, it is beneficial for a QKD network to avoid those nodes not considered sufficiently reliable in a trusted network topology. Those "untrusted" nodes can be easily bypassed by defining an alternative route on the overlay level.

In [27], the efficiency of an overlay technique which aimed to reduce overhead in a QKD network was investigated through the use of jumbo frames[1] on the overlay layer and whose fragmentation on the underlay TCP/IP stack would reduce the number of headers to be encrypted (Fig. 4.14). In this manner, the freedom to select or use multiple overlay routes can be maintained, and the cost to protect additional headers can be minimized.

A simulation experiment using QKD Network Simulator (QKDNetSim) showed that the approach described above is justified in some cases since it achieves greater efficiency and resilience to network outages. However, overlay network performance may be compromised if different TCP protocol combinations are

---

[1] In data networking, the standard Ethernet Maximum Transmission Unit (MTU) is 1500 bytes, and frames with higher MTUs are generally known as jumbo frames.

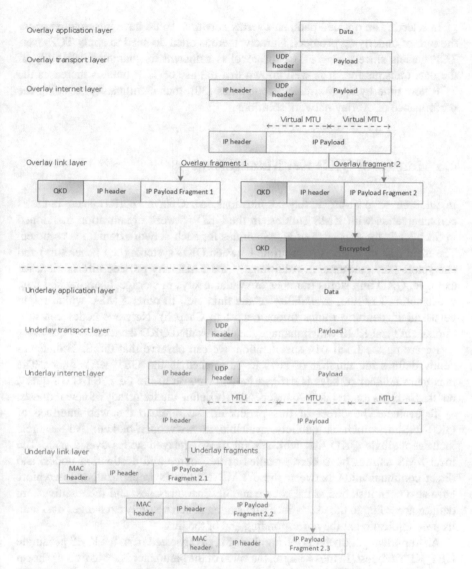

Fig. 4.14 Example of fragmentation of a packet induced by an encapsulation layer. Suppose user packets are encrypted using OTP. A dedicated QKD header is used to transmit OTP details (keyID, sessionID, etc.). Each overlay fragment contains an IP header which is encrypted with the user's data on the underlay stack, thereby increasing the network overhead

applied on the overlay and underlay network stacks because of the occurrence of the TCP "meltdown problem"[2] [28], TCP proxy [29] or other specific TCP phenomena.

_____

[2] The meltdown occurs when TCP is applied on an overlay or a lower stack which has a disagreement between negotiation of the TCP timer and the Maximum Segment Size (MSS)

In selecting an optimal path, an overlay network should have information about the type of underlying protocol. Namely, there is often no need to apply TCP-over-TCP tunnels since a single TCP protocol is sufficient to guarantee reliability of the data transmission. It is well known that the use of TCP tunnels increases the TCP flow time by approximately four times [30], thus significantly reducing the performance of overlay network solutions.

4.3   Impact of QKD Key Management

In addition to post-processing applications, MAC layer performance includes communication with KMS entities. In the QKD network organization mentioned in Sect. 3.1.2, the existence of KMS entities for each network domain is assumed. The main task of KMS is to arbitrate between QKD systems (key generators) and encryptors (key consumers). Additionally, the KMS may provide information, for example, QKD link status (number of available keys in storage, charging key rate, consumption key rate, availability of the link, etc.) to other KMSs, which might be useful in defining routes (more details in Chap. 6). Network nodes can also implement local KMS which manage locally installed QKD devices.

Regarding the ETSI 014 specification, we can observe that the SAE independently defines the number of keys to be fetched in the GET_KEY query. The maximum number of keys is defined in the response to the GET_STATUS query, but the SAE has the freedom to independently define the frequency of new requests.

To evaluate the effect of this parameter, we refer to the web interface of QKDNetSim, which is publicly available at www.open-qkd.eu. An example includes a single QKD link with several trusted relayed nodes over a path. The local KMS entities have been installed at the source and destination nodes, and direct communication between these KMS entities is implemented. Encryptors have also been installed at the source and destination nodes, and their settings are defined according to the user's SAE application which securely exchanges data with its peer application at the corresponding end of the link.

An important parameter is the number of keys requested from the KMS per single GET_KEY request. In the example, the value of the parameter is set to five, as shown in Fig. 4.15. The SAE will use the same key to encrypt five packets with an AES cipher and then request additional keys to encrypt the next five packets, and so on. Figure 4.16 shows the exchange of 249 packets between SAE applications and that

---

values. If the timeout for receiving acknowledgment of a previously sent fragment expires, the underlying TCP protocol will reschedule retransmissions and increase the timeout interval so that the connection is not interrupted. However, the TCP overlay connection is not aware of this and continues to retransmit packets more quickly than the subsequent level, thus increasing the congestion. This further slows the connection, which then prompts needless retransmissions on the overlay layer. The upper layer generates more data than the lower layer can handle, which can diminish or even stop a connection [28].

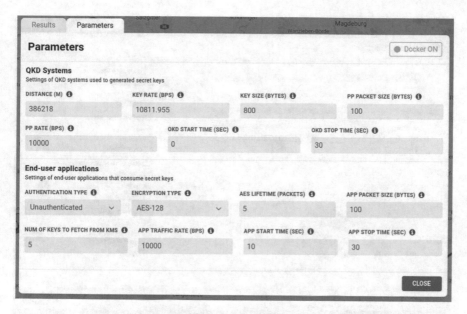

Fig. 4.15 Configuration of QKDNetSim web interface parameters available at www.open-qkd.eu

21 packets were exchanged between SAE applications and KMS. If the parameter "number of keys to fetch from KMS" is increased to 20, lower values are obtained, as shown in Fig. 4.17.

A comparison of these diagrams reveals that the gradients of the red curves which describe the intensity of keys served to applications are not the same. In the second case, the curve has a stepped shape with a shallower gradient, which means that application periodically requires a larger number of keys. Consequently, the number of packets exchanged with the KMS is reduced. For key storage, the application implements a local buffer, which is periodically refreshed with new keys from KMS. Comparing performance, the application exchanges more packets with the corresponding application than the case when no local buffer is implemented.

Table 4.3 lists the results of simulations with different cryptographic methods applied. The signaling packets are those packets exchanged between source and destination nodes which exchange keyIDs information. This communication is beyond the scope of ETSI 014, but it is necessary to apply GET_KEY_WITH_KEYIDS to fetch the same set of keys from KMS. The column "# keys consumed" lists the number of keys spent from the KMS perspective. The number of keys delivered to the user and the number of keys actually used for encryption may differ because of the existence of local buffers and the application's encryption policy. This can result in different numbers of exchanged packets and consumed keys when OTP is applied.

By applying AES, a key 128 bits long is used in the encryption of multiple packets. However, when OTP is used, the KMS needs to deliver a larger number

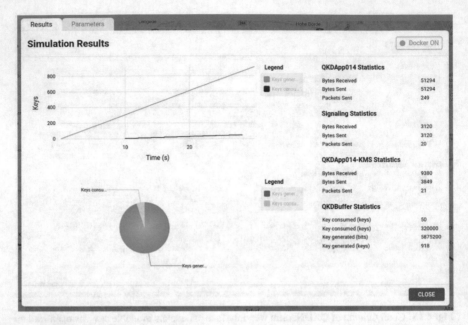

Fig. 4.16 Results of QKDNetSim simulations with the "Number Of Keys To Fetch from KMS" parameter set to a value of 5

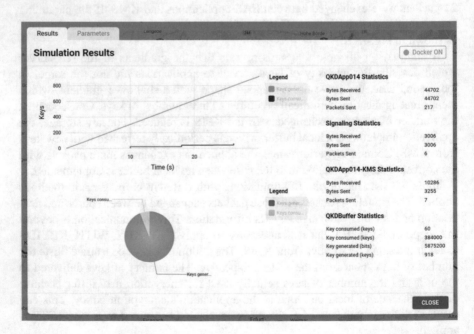

Fig. 4.17 Results of QKDNetSim simulations with the "Number Of Keys To Fetch from KMS" parameter set to a value of 20

Table 4.3 Comparison of the traffic between QKD entities when different cryptographic methods are applied

Cryptographic method	# data packets	# signaling data packets	# packets to KMS	# keys consumed	# keys generated
Unprotected	167	0	0	0	586
OTP	161	34	35	170	586
OTP + VMAC	60	13	15	70	586
AES (*refresh 5)	164	8	9	40	586
AES (*refresh 5) + VMAC	166	8	9	40	586
AES (*refresh 10)	167	4	v5	20	586
AES (*refresh 10) + VMAC	167	4	5	20	586

of keys and supply them more frequently. Therefore, when the application requests a large number of keys (defined by the parameter "number of keys to fetch from KMS"), it will wait for a longer time[3] since more time is required to generate the requested number of keys. Additionally, IP fragmentation or specific TCP phenomena such as "TCP silly window syndrome" avoidance algorithms increase the delay in delivering keys from KMS to the application.

The example above requires explanation of the application's behavior from the aspect of KMS overload protection. If the application sends a GET_KEY request and receives a negative response from KMS with the information that an insufficient number of keys is available, the application will be "punished" to attempt again after three seconds (Fig. 4.18). This type of behavior is a detail of QKDNetSim implementation and can be changed in the settings. It is introduced to protect KMS entities from Distributed Denial-of-Service (DDoS) attacks by malicious applications. Therefore, if we consider an example with empty QKD key buffers, and if the application requests keys more frequently, it will be punished more frequently and spend more time waiting. This behavior occurs mostly when OTP or AES with a high refresh rate is used. Table 4.4 lists the performance for different application rates.

All the above shows that the effect of KMS communication on MAC layer performance is significant. In the example, only KMSs which are directly related are included. The delay in serving keys is even more apparent when multiple KMSs are included in the path. The results also show that the application's design has an important effect on performance. The ETSI 004 and 014 specifications define how the application and KMS communicate, although the choice of communications settings (how frequently and when a key is requested) is given to the designer of the final application.

---

[3] The application will not be able to continue encryption of the user's data because of the lack of available key material. It will therefore "wait" until available key material is obtained from KMS.

ETSI QKD 014 - Protocol and data format of REST-based key delivery API

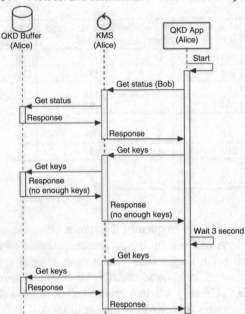

Fig. 4.18 When an insufficient number of keys is available, the application waits three seconds before again requesting keys from KMS. This is a setting in QKDNetSim which reduces malicious activity

Table 4.4 Comparison of performance for different application rates

Application rate [kbps]	Encryption method	# data packets exchanged	# keys consumed
10	Unencrypted	167	0
	OTP	161	170
	AES (10)	167	20
20	Unencrypted	334	0
	OTP	280	280
	AES (10)	328	40
50	Unencrypted	1667	0
	OTP	130	140
	AES (10)	795	80
100	Unencrypted	1667	0
	OTP	140	150
	AES (1)	270	270
	AES (10)	1486	150

4.4 Summary

This chapter described the performance of the lower layers of the QKD stack, which according to the common organization of IP networks may be considered a single MAC layer. The stack includes operations with quantum optical equipment, post-processing communications, and key-relay and key management functionalities. Each of these components can significantly affect performance, primarily in the rate of cryptographic key generation. Applications above the MAC layer are generally not concerned with how keys are generated and treat this layer as a service. The MAC layer should ensure that applications have sufficient keys for smooth operation, although how this is implemented is not relevant issue to the application layer. The quality of service at this layer includes a large number of settings and parameters which define the type of QKD protocol and type and quality of media to create the quantum channel. This also includes organization of the post-processing module and ultimately the manner of serving or organizing keys through KMS communication.

Although applications generally do not consider QKD MAC layer settings, it is important to note that the MAC layer frequently seeks to protect its resources from malign applications. Excessive key requests may be characterized as malicious activity and be blocked to maintain fairness or the efficient use of resources.

References

1. Kumar, R., Qin, H., & Alléaume, R. (2015). Coexistence of continuous variable QKD with intense DWDM classical channels. *New Journal of Physics, 17*, 043027. ISSN: 13672630. https://doi.org/10.1088/1367-2630/17/4/043027
2. Mao, Y., Wang, B.-X., Zhao, C., Wang, G., Wang, R., Wang, H., Zhou, F., Nie, J., Chen, Q., Zhao, Y., Zhang, Q., Zhang, J., Chen, T.-Y., & Pan, J.-W. (2018). Integrating quantum key distribution with classical communications in backbone fiber network. *Optics Express, 26*(5), 6010. ISSN 1094-4087. https://doi.org/10.1364/OE.26.006010.
3. Zavitsanos, D., Ntanos, A., Giannoulis, G., Avramopoulos, H. (2020). On the QKD Integration in Converged Fiber/Wireless Topologies for Secured, Low-Latency 5G/B5G Fronthaul. *Applied Sciences, 10*(15), 5193. ISSN 2076-3417. https://doi.org/10.3390/app10155193
4. Alléaume, R., Branciard, C., Bouda, J., Debuisschert, T., Dianati, M., Gisin, N., Godfrey, M., Grangier, P., Länger, T., Lütkenhaus, N., Monyk, C., Painchault, P., Peev, M., Poppe, A., Pornin, T., Rarity, J., Renner, R., Ribordy, G., Riguidel, M., . . . & Zeilinger, A. (2014). Using quantum key distribution for cryptographic purposes: A survey. *Theoretical Computer Science, 560*(P1), 62–81. ISSN 03043975. https://doi.org/10.1016/j.tcs.2014.09.018
5. Gyongyosi, L., Imre, S., & Nguyen, H. V. (2018). A survey on quantum channel capacities. *IEEE Communications Surveys and Tutorials, 20*(2), 1149–1205. ISSN 1553-877X. https://doi.org/10.1109/COMST.2017.2786748
6. Scarani, V., Bechmann-Pasquinucci, H., Cerf, N. J., Dušek, M., Lütkenhaus, N., & Peev, M. (2009). The security of practical quantum key distribution. *Reviews of Modern Physics, 81*(3), 1301–1350. ISSN 0034-6861. https://doi.org/10.1103/RevModPhys.81.1301

7. Sharma, P., Agrawal, A., Bhatia, V., Prakash, S., Mishra, A. K. (2021). Quantum key distribution secured optical networks: A survey. *IEEE Open Journal of the Communications Society, 2*(July), 2049–2083. ISSN 2644-125X. https://doi.org/10.1109/OJCOMS.2021. 3106659

8. Wolf, R. (2021). *Quantum key distribution*. Lecture Notes in Physics (Vol. 988). Cham: Springer International Publishing. ISBN 978-3-030-73990-4. https://doi.org/10.1007/978-3-030-73991-1

9. Maurhart, O., Pacher, C., Happe, A., Lor, T., Tamas, C., Poppe, A., & Peev, M. (2013). New release of an open source QKD software: design and implementation of new algorithms, modularization and integration with IPSec. In *Qcrypt 2013*.

10. Mehic, M., Maurhart, O., Rass, S., Komosny, D., Rezac, F., & Voznak, M. (2017). Analysis of the public channel of quantum key distribution link. *IEEE Journal of Quantum Electronics, 53*(5), 1–8. ISSN 0018-9197. https://doi.org/10.1109/JQE.2017.2740426

11. Lessiak, K., Kollmitzer, C., Schauer, S., Pilz, J., & Rass, S. (2009). Statistical analysis of QKD networks in real-life environments. In *Proceedings of the Third International Conference on Quantum, Nano and Micro Technologies* (pp. 109–114). IEEE Computer Society. https://doi.org/10.1109/ICQNM.2009.30

12. Rass, S. (2009). Simple error scattering model for improved information reconciliation. *Preprint arXiv:0908.2069*.

13. Rass, S., Kollmitzer, C., 2010. Adaptive Cascade, in: Applied Quantum Cryptography. Springer, Lecture Notes in Physics (LNP, volume 797), pp. 49–69, ISBN: 978-3-642-04831-9.

14. Rass, S., & Schartner, P. (2010). Non-interactive information reconciliation for quantum key distribution. In *Proceedings of the 24th IEEE International Conference on Advanced Information Networking and Applications* (pp. 1054–1060). IEEE Computer Society Press. https://doi.org/10.1109/AINA.2010.27

15. Rass, S., Schartner, P., & Greiler, M. (2009). Quantum coin-flipping-based authentication. In *ICC 2009 Communication and Information Systems Security Symposium*, Dresden. https://doi.org/10.1109/ICC.2009.5199383

16. Berlin, G., Brassard, G., Bussiéres, F., & Godbout, N. (2008). Loss-tolerant quantum coin flipping. In D. Avis, C. Kollmitzer, & V. Ovchinnikov (Eds.), Second international conference on quantum, Nano and Micro technologies (ICQNM 2008) (p. 1–9). IEEE. https://doi.org/10. 1109/ICQNM.2008.17

17. Bennett, C. H., & Brassard, G. (1984). Quantum cryptography: Public key distribution and coin tossing. In *Proceedings of IEEE International Conference on Computers, Systems and Signal Processing* (Vol. 175, pp. 8). New York: Elsevier B.V. https://doi.org/10.1016/j.tcs.2011.08. 039

18. Rass, S., König, S., & Schauer, S. (2015). BB84 quantum key distribution with intrinsic authentication. In *Proceedings of Ninth International Conference on Quantum, Nano/Bio, and Micro Technologies (ICQNM)* (pp. 41–44). ISBN: 978-1-61208-431-2

19. Hares, S., Rekhter, Y., & Li, T. (2006). A border gateway protocol 4 (BGP-4). *RFC 4271*. ISSN 1098-6596. https://doi.org/10.1017/CBO9781107415324.004

20. Labovitz, C., Ahuja, A., Wattenhofer, R., & Venkatachary, S. (2001). The impact of Internet policy and topology on delayed routing convergence. In *Proceedings IEEE INFOCOM 2001. Conference on Computer Communications. Twentieth Annual Joint Conference of the IEEE Computer and Communications Society (Cat. No.01CH37213)* (Vol. 1, pp. 537–546). ISSN 0743-166X. https://doi.org/10.1109/INFCOM.2001.916775

21. Augustin, B., Friedman, T., & Teixeira, R. (2011). Measuring multipath routing in the internet. *IEEE/ACM Transactions on Networking, 19*(3), 830–840. ISSN 10636692. https://doi.org/10. 1109/TNET.2010.2096232

22. Teixeira, R., Marzullo, K., Savage, S., & Voelker, G. M.(2003). Characterizing and measuring path diversity of internet topologies. *SIGMETRICS Performance Evaluation Review, 31*(1), 304–305. ISSN 0163-5999. https://doi.org/10.1145/781064.781069

23. Begen, A. C., Altunbasak, Y., Ergun, O., & Ammar, M. H. (2005). Multi-path selection for multiple description video streaming over overlay networks. *EURASIP Journal of Signal Processing: Image Communications, 20*(1), 39–60. ISSN 09235965. https://doi.org/10.1016/j.image.2004.09.002

24. Tang, C., & McKinley, P. K. (2005). Improving multipath reliability in topology-aware overlay networks. In *25th IEEE International Conference on Distributed Computing Systems* (pp. 82–88).

25. Tao, S., Xu, K., Estepa, A., Fei, T., Gao, L., Guerin, R., Kurose, J., Towsley, D., & Zhang, Z. L. (2005). Improving VoIP quality through path switching. *Proceedings - IEEE INFOCOM, 4*(C), 2268–2278. ISSN 0743166X. https://doi.org/10.1109/INFCOM.2005.1498514

26. Ma, Z., Shao, H.-R., & Shen, C. (2004). A new multi-path selection scheme for video streaming on overlay networks. In *2004 IEEE International Conference on Communications (IEEE Cat. No.04CH37577)* (Vol. 3, pp. 1330–1334). IEEE. ISBN 0-7803-8533-0. https://doi.org/10.1109/ICC.2004.1312728

27. Mehic, M., Komosny, D., Mauhart, O., Voznak, M., & Rozhon, J. (2016). Impact of packet size variation in overlay quantum key distribution network. In *2016 XI International Symposium on Telecommunications (BIHTEL)* (pp. 1–6). Sarajevo, Bosnia and Herzegovina. IEEE. ISBN 978-1-5090-2902-0. https://doi.org/10.1109/BIHTEL.2016.7775711

28. Honda, O., Ohsaki, H., Imase, M., Ishizuka, M., & Murayama, J. (2005). Understanding TCP over TCP: Effects of TCP tunneling on end-to-end throughput and latency. In M. Atiquzzaman & S. I. Balandin (Eds.), *Optics East* (p. 60110H). https://doi.org/10.1117/12.630496

29. Maki, I., Hasegawa, G., Murata, M., & Murase, T. (2005). Performance analysis and improvement of TCP proxy mechanism in TCP overlay networks. In *IEEE International Conference on Communications, 2005. ICC 2005. 2005* (Vol. 1, pp. 184–190). IEEE. ISBN 0-7803-8938-7. https://doi.org/10.1109/ICC.2005.1494344

30. Lee, B. P., Balan, R. K., Jacob, L., Winston, W. K. G., Seah, K. G., & Ananda, A. L. (2002). Avoiding congestion collapse on the Internet using TCP tunnels. *Computer Networks, 39*(2), 207–219. ISSN 13891286. https://doi.org/10.1016/S1389-1286(01)00311-5

Chapter 5
Quality of Service Signaling Protocols in Quantum Key Distribution Networks

Generally, the need for and the type of signaling, or more strictly speaking, the QoS signaling protocol depends on the applied QoS architecture. Here, we distinguish between the *signal* and the *signaling* protocol, which in the literature elsewhere are commonly applied as synonyms.

The signal protocol is the mechanism which enables the transmission of information concerning changes occurring within the network. This information is often encoded in binary form to reduce network overhead. Examples include Explicit Congestion Notification (ECN), header flags in detected routing loops, notifications for routing mechanisms (more details in Sect. 6.5) and others. On the other hand, QoS signaling protocols have a more complex task involving the exchange of information about required QoS parameters and the allocated/reserved network resources for specified network flow. Examples include reservation of network resources, setting firewall pinholes and network address bindings (NAT), and exchanging route diagnostic information. They are also used for the interchange of information between nodes to create, maintain, renegotiate and finally tear down network sessions. In the context of this chapter, signaling protocols are of primary interest.

It is important to highlight that signals and signaling protocols do not define the route within the network; this task is given to routing protocols. Routing protocols can take advantage of the information exchanged via signals/signaling protocols, for example, to ignore paths which do not satisfy the requested QoS specifications as a result of congestion or other problems.

Although modern IP networks allow packets to follow arbitrary routes to a destination, in practice, network management is based on flow organization which mainly uses single route. Let us suppose that the exchange of traffic between nodes located at different ends of a network is attained through multiple network routes. Signaling information concerning congestion and the collection of statistics on network links would not provide any gains given that the validity of the obtained data would be inaccurate. Consider also an end-to-end link exchange between two

© Springer Nature Switzerland AG 2022
M. Mehic et al., *Quantum Key Distribution Networks*,
https://doi.org/10.1007/978-3-031-06608-5_5

distant nodes: if each packet is indeed delivered via a different route, calculation of the critical parameter of the most commonly used transport TCP protocol would not be up to date. Deviations in the measured Round-Trip Time (RTT) values[1] would be significant and not useful for the functionality of the TCP protocol (defining the size of TCP windows, determining the length of queues, retransmission timers, etc.) [1]. Determining an arbitrary route for each packet would also imply a violation of consistent routing, which is contrary to per-hop-behavior approaches and the basic postulates of routing protocol design [2–4]. In practice therefore, all traffic follows the same route to the destination. When the quality of communication over the route is disrupted, it is then necessary to propagate information concerning the event that occurred and use an alternative route.

The exchange of signaling messages is primarily performed over the route defined by the routing protocol. If the allocation and reservation mechanisms of network resources are taken into account, it would not make sense to reserve resources on the path over which the user's data will not be routed. Signaling packets therefore indicate the arrival of data packets over the same route on which they are exchanged and on which they seek to provide available QoS resources.

In knowing the difference between signals and signaling mechanisms, we can identify the basic requirements for efficient use of these solutions in QKD networks. A signaling protocol well-suited for operation in a QKD network should fulfill the main design objectives listed by priority, as follows:

- Given the importance of signals/signaling messages and the significance of their potential misuse, it is necessary to at least to authenticate communications and use encryption whenever possible. If an attacker is able to intercept and modify the exchanged values, he could influence the selection of the route subsequently defined by the routing protocol. More precisely, he could direct the traffic over those nodes under his control [5].
- Because the exchange of these signaling messages must be protected, it is necessary to reduce their number. Extensive communications implies the consumption of cryptographic keys to protect service information instead of protecting data traffic [6].
- The information exchange mechanism should be fast and timely. Those mechanisms that depend on the exchanged information (i.e., routing protocols) can then respond quickly to unexpected network events.

Taking into account the transmission information technique, signaling protocols can be categorized into in-band and out-of-band solutions. The former includes protocols which combine control and data in the same traffic, while the latter refers to the technique applied to provide control information via explicit control packets.

---

[1] RTT indicates the amount of time it takes for a message to be delivered from the source to the destination, with confirmation of delivery from the destination to the source.

5.1 In-Band signaling and QKD

The QKD Signaling Protocol (QSIP) reported in [7] is the first QKD in-band signaling protocol. It was inspired by the similarity of QKD and ad hoc networks, or more precisely, the INSIGNIA in-band signaling protocol. The primary aim of QSIP is to share information without reserving resources. QSIP should enable changes in session settings and addresses the following questions: how do users agree on which cryptographic method to use? How do they change the selected cryptographic method during a session (i.e., switching from AES-256 to OTP)? How is a session terminated and how are resources freed?

Before considering the details of the QSIP protocol, we briefly review the INSIGNIA functionalities. The INSIGNIA protocol was designed for per-flow end-to-end reservation of network resources. Instead of sending dedicated signaling packets which are potentially unreliable due to the network dynamics, INSIGNIA uses the OPTION field in the IP header (which is then referred to as INSIGNIA OPTION) to transmit additional information about the required QoS resources [8]. The signaling packets are transmitted together with user traffic to explore the availability of a path during data transmission [9].

The information concerning the flow state is processed in a soft-state, that is, the flow state information is regularly updated. In cooperation with the admission controller, INSIGNIA provides network resources (bandwidth) to the flow when demands can be fulfilled. Otherwise, flow is adopted to the best possible service but without producing signals of rejection or error. Intermediate nodes that receive packets with degraded QoS information will not relocate resources, and the flow reservation is not refreshed. When the reservation terminates, resources are deallocated. However, when a package with a valid QoS indication is received, soft-state times are updated. Finally, when the packet reaches the destination, a dedicated report is sent to the source node stating which links meet the required QoS requirements and which can be used for further communication.

5.1.1 QSIP: A Quantum Key Distribution Signaling Protocol

QSIP aims to support QKD multicast communication. To attain this goal, all participants must share the same session identifier. Instead of using the popular 5-tuple flow identifier which contains multiple values (IP source address, IP destination address, transport protocol identifier, source port, and destination port), QSIP uses a session identifier created as a random value to uniquely identify a signaling session. A session can be mapped to a certain flow, but it can provide flexible relationships such as multihoming and IPv4/v6 traversal. Therefore, the session is not strictly bound to IP addresses and can involve multiple network flows, allowing simple use of QSIP to establish group multicast communications. Although the method of specifying the QSIP session identifier is not described, it can be viewed as a KSID through the prism of the ETSI 004 and ETSI 014 specifications.

QSIP has also been inspired by the well-known Session Initiation Protocol (SIP) protocol [10, 11]. It follows the $\langle type \rangle = \langle value \rangle$ message format. QSIP describes two message types:

- QUERY: this message is used to initiate the session by transmitting information about cryptographic algorithms supported by the source node. Since the message is transmitted along with data, it aims to examine possible encryption and authentication algorithms supported by all nodes on the route to the destination. This is necessary, for example, if the QKD network does not follow a key-relay approach, that is, each packet is literally transmitted with a hop-by-hop approach to the destination (Fig. 2.5). For example, Alice can define supported cryptographic methods by specifying encryption $e \in \{AES256, AES128, OTP, IDEA128\}$ and authentication $a \in \{VMAC, KRAWCZYK, BIERBRAUER\}$ values. Additional options are possible, but at any time Alice only supports those listed in parentheses. It is evident that QUERY can also convey accompanying information, such as the number of keys, to request from the KMS when an ETSI 014 GET_KEY query, key refresh rate or other query is sent.
- ACK: this implies feedback on the selected cryptographic tool. The intermediate nodes on the path and destination node will analyze the supported options and respond with an ACK message to inform the node that the session has been accepted, specifying the supported encryption algorithms, i.e., $e = AES126$ and authentication $a = VMAC$. Since QSIP is an in-band protocol, an ACK message cannot be sent immediately. It is stored in the local buffer and waits for the first data packet to be routed to the source node and convey the ACK response. If the same source has additional QUERY requests in the meantime, they are ignored until the previously processed request is fully processed by freeing memory to store the new ACK message. If no supported algorithms are available, then empty values for e and a are returned.

QSIP has a primary application in the distributed QKD networks discussed in Chap. 6. Namely, no centralized management is performed in those networks, but the route is established dynamically. Since QSIP transmits values along with data, efforts are made to reduce key consumption by directing packets to those links with sufficient resources. The intermediate node may respond with an empty ACK before the packet arrives at the destination node, thus reducing key consumption. Combined with caching mechanisms, this approach can notably improve performance (Sect. 6.5.2). Each QSIP message also includes a time stamp value indicating the time when the packet was sent. The transmission delay can then be evaluated, which may be of interest to the routing protocol when it defines the optimal route.

After a session is established, QSIP can be used to modify the settings of the ongoing session. In the same manner as the procedure described above, the source or destination node can start a QUERY procedure which will be responded to with an ACK message. Since QSIP is not a flow-oriented reservation protocol, there are

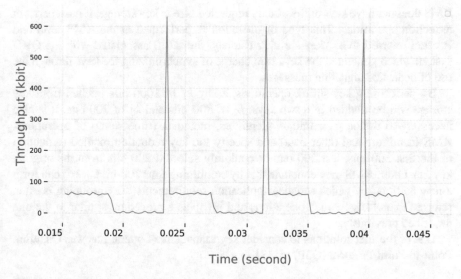

Fig. 5.1 The throughput of an authentication module in the AIT R10 QKD post-processing application. Periodic repetition of the "heart-rate" pattern begins with the exchange of sifting information and error-reconciliation processing which dominates the overall amount of network traffic

no reservations which need to be released. QSIP does not therefore implement any specific tear-down mechanisms.

QSIP is designed to take advantage of already authenticated packets within post-processing traffic. Detailed traffic analysis of the AIT R10 QKD post-processing software [12] showed that at least 14 kbps of traffic generated by the authentication module is always present in the QKD link (Fig. 5.1). Since these packets have already been authenticated, QSIP suggests extending the authenticated packet payloads with routing or signaling information. It is an elegant approach in implementing the distribution of routing and signaling packets without adding overhead traffic. We may note that it also represents a potential solution for the transmission of KMS–KMS related data. Regarding the positioning of QSIP values in packet headers, the inclusion of signaling data in the specific QKD (extended Q3P) headers is a proposition we discuss further below.

5.2   Out-of-Band Signaling and QKD

Besides the information necessary to establish, maintain and terminate a session, the concept of signaling in QKD networks includes additional functionalities. One of these tasks is signaling the management and organization of cryptographic keys.

Consider an example in which QKD systems generate 1000-bit keys. The application requires KMS to provide 800-bit keys for OTP encryption. Since the

KMS does not have keys of the exactly requested size in key storage, it must perform reduction operations. This type of operation is performed at the KMS level and is often referred to as "key-shrink". Namely, the KMS associated with the QKD system which generated the keys is in charge of synchronizing the keys through the use of dedicated signaling messages.

By performing key-shrink operations, a key k_1 of 1000 bits fetched from key storage will be divided into two keys: k_2 of 800 bits and k_3 of 200 bits. It is also necessary to define their unique identifiers, metadata (time stamp of operations, KMS identifiers and other data) and specify the key reduction method (reduction of the first 200 bits, last 200 bits or randomly selected 200 bits from the original key k_1). Later, KMS may conclude that by combining four 200-bit keys, it can form a new 800-bit key which satisfies application requirements. Key grouping is often referred to as "key-merge" operations and involves a procedure similar to the one described previously.

One of the first solutions to consider key management operations was Quantum Point-to-Point Protocol (Q3P) [6, 13].

5.2.1   Q3P: Quantum Point-to-Point Protocol

Since QKD systems operate in point-to-point mode, the inspiration for Q3P was the Point-to-Point Protocol (PPP) operating at the second TCP/IP layer [14]. The authors maintained that the second layer is the most the convenient for performing security operations since no additional encapsulation is required [13].

Q3P is conceived as a comprehensive framework for the implementation of all operations related to the implementation of QKD networks. Although initially intended as a link between an application which requires keys at the top of the TCP/IP framework and QKD devices that generate keys, Q3P defined additional requirements from the application layer to avoid network congestion.

5.2.1.1   Q3P Key Store

The main motivation for the Q3P framework is efficient key management. Since the rate of key generation is low, efficient tools are required to reduce the unnecessary consumption of keys to a minimum. Special attention is given to the organization of key storage (cf. [15, 16]). Primarily, a master/slave arrangement is introduced in which the master party dictates the key storage organization. To avoid collisions, keys are organized in Pickup Stores, Common Stores and I/O buffers; see Fig. 5.2 [17].

All keys generated by QKD systems are stored in the Pickup Store. However, before the keys are used, synchronization must be performed (confirmation that the same material is found in both Pickup Stores of the nodes which form the corresponding QKD link). While the keys are held in the Pickup Store, they contain

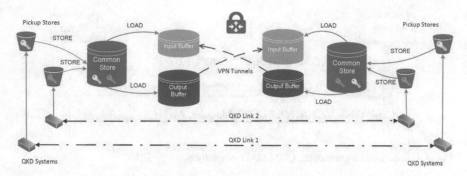

Fig. 5.2   Q3P key storage architecture

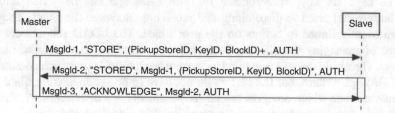

Fig. 5.3   Sequence diagram of the Q3P STORE subprotocol

metadata (keyID, time stamp, error rate in the generation of the key, epsilon security, type of device which produced the key, and other details) which can be used for filtering and synchronization. When the keys are eventually filtered and synchronized, they are organized into blocks of bits containing no metadata and moved into the Common Store. These blocks can have different sizes and be accessed using BlockIDs.

The Q3P link has only one Common Store (which may contain one or more QKD links between connected nodes). Unlike other repositories, the Common Store is physically implemented in memory, which allows long-term storage of synchronized keys. If the QKD system experiences an outage, Pickup Stores and I/O buffers can be reset (discarding all previously stored keys). The Common Store maintains storage of the keys in hardware, thereby avoiding additional resynchronization. To migrate the keys from the Pickup Store to the Common Store, the STORE subprotocol can be applied.

Figure 5.3 shows that all messages are authenticated. The initial message MsgId-1 contains the identifiers for the Pickup Store, key and the block which will be created in the Pickup Store at the master node. Upon receiving the message, the slave prepares its key in the Pickup Store for the Common Store and responds with the confirmation message MsgId. The master then pushes its key into the Common Store and responds with the acknowledgment that the keys are synchronized and ready for use. Upon receiving acknowledgment, the slave will push its key and complete the STORE subprotocol operations.

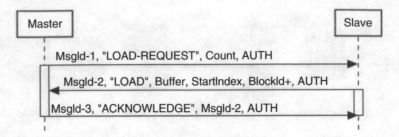

Fig. 5.4 Sequence diagram of the Q3P LOAD subprotocol

The keys are now synchronized on both sides and can be sorted into the I/O buffers dedicated to dispatching and receiving encrypted messages. They are symmetrically linked to buffers on the peer's side. The LOAD subprotocol is in charge of managing the allocation of keys from the Common Store to the I/O buffers. Threshold values are applied to warn of the lack of keys, which commences the LOAD procedure, but I/O buffers do not have to be symmetrically filled. It is the master node which analyzes the key requirements and determines whether the communication is simplex (in one direction), resulting in asymmetric key allocation, or duplex (in both directions), which will indicate symmetric allocation in the I/O buffers.

In Fig. 5.4, the procedure of the LOAD subprotocol is shown to be similar to the STORE subprotocol, and we can observe that all the messages are authenticated. Because the subprotocols are continuously executed, we may be concerned that spending keys on message authentication will lead to a reduction in the number of available keys. Delayed authentication could provide a single authentication check for multiple message interactions between nodes. Although this represents a potential solution to increased key consumption, in practice it raise issues around practical implementation and implies that messages must be stored in memory and that follow-up actions after successful authentication cannot be performed instantly. If authentication is unsuccessful, it is necessary to discard the scheduled operations and revert the execution of synchronization procedures to those steps in which the last successful authentication was performed. In practice, this represents multiple challenges in terms of memory management.

Q3P proposed the implementation of Q3P headers which convey information about the STORE and LOAD protocols. The header shown in Fig. 5.5 was later identified as a potential solution for the transfer of encrypted data content, especially in cases when OTP encryption is applied.

We may note the existence of a "length" field which indicates the packet size in the case of applied encryption. This field has an important function in deserializing decrypted packets, i.e., unpacking the header of the higher layers after decryption has been performed. Q3P includes the crypto engine which performs encryption and authentication operations. It is used to compute different authentications with

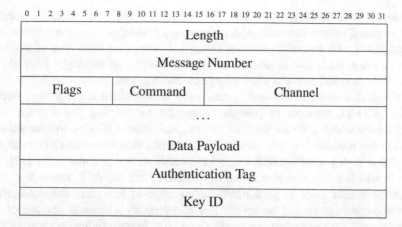

Fig. 5.5 Structure of Q3P header

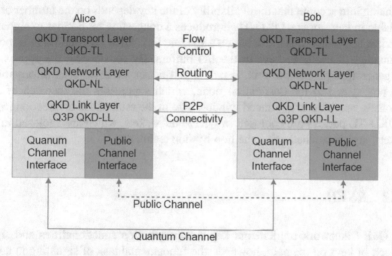

Fig. 5.6 The Q3P protocol stack

various authentication tag sizes in parallel to provide the crypto context for the established communication channel.

5.2.1.2   Q3P Network Stack

The inspiration for Q3P was the PPP protocol, and it was initially designed to work on a second layer. However, the Q3P framework was also proposed to encompass solutions for higher TCP/IP network layers (Fig. 5.6).

On the second layer, the Q3P protocol operates with extended PPP functionalities. On the third layer, Q3P implements the QKD-NL solution, which was developed from the IP protocol and OSPFv2 routing [18]. The QKD-TL solution was proposed for the transport layer, which was developed from the popular TCP protocol, but with several significant changes. Namely, the basic task of this layer is to ensure reliable and sequential end-to-end delivery of messages relayed over re-transmission and return acknowledgments mechanisms.

In a packet-switched network, packets are retransmitted if congestion occurs, while QKD-TL attempts to prevent congestion by reacting pro-actively, using mechanisms which examine the state of links and nodes. QKD-TL bypasses the re-sending mechanisms implemented in TCP, asserting that the content of the packets in QKD is highly sensitive with respect to confidentiality or authenticity [19]. The case in which a message is successfully delivered but the ACK message did not reach the source node is particularly noteworthy: in this case, the successfully delivered message should be resent and consume an additional amount of key material, and the re-sending mechanism may also impose limits on re-using keys from I/O buffers.

Taking into account that the availability of the key depends on the number of keys available in key storage, QKD-TL introduces a congestion bit notifier to signal that sufficient key material is available on a specific link. Each host should also monitor the number of available keys in the I/O buffers. By considering the approximate key generation rate of the link, it can signal a lack of keys. This information is then propagated to the destination node, which suppresses the dispatch of ACK responses, resulting in a reduced window size in the established connection [6].

QKD-TL packets are enclosed in QKD-NL packets, which are enclosed in Q3P packets, and dispatched using the hop-by-hop technique.

5.2.2   RSVP

The Q3P framework implements key synchronization functionalities and signals the lack of keys on the path, however, the fundamental task of signaling in a QKD network is negotiation of the key which will be used for cryptographic operations.

To guarantee the existence of the required number of keys and ensure their availability, some solutions turn to reserving resources in advance. For those purposes, the RSVP protocol, which has been successfully used in IP networks for many years, is usually applied [20]. Quantum Key Reservation Approach (QKRA) falls under the category of IntServ QoS models (Sect. 3.1) and is based on the reservation of keys on intermediate nodes along the path from the source node to the destination node. A report on the possibility of reserving keys is submitted to the destination node, which summarizes the status information and evaluates the reservation of available resources. Finally, the source node is informed of successful reservations on intermediate nodes [21].

The most recent approach, however, is the use of RSVP in MPLS or Generalized Multi-Protocol Label Switching (GMPLS) environments to automate the key synchronization process. It is used to exchange encryption requirements and

	0 1 2 3 4 5 6 7 8 9 10 11 12 13 14 15 16 17 18 19 20 21 22 23 24 25 26 27 28 29 30 31		
L	Type	Length	Session ID
	··· Session ID ···		
	Session ID		Key Length
Enc_layer	Ref Type		Refresh value
	Enc_Algo		

Fig. 5.7 Structure of RSVP ERO subobject and its key and encryption details [22]

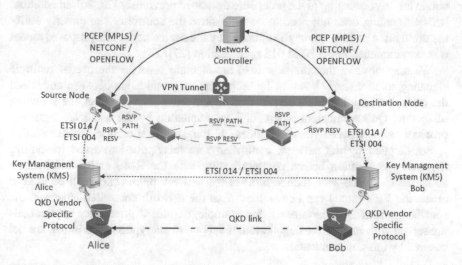

Fig. 5.8 Schematic view of signaling solutions applied to communications with QKD entities

collect information about the success of key reservation. As shown in Fig. 5.7, dedicated Explicit Route Object (ERO) subobject has been proposed to include details of the key length, encryption algorithm, key refresh values and others [23], i.e., RSVP is used to convey information about the key, but key reservation and acquisition is performed using ETSI 004 [24]. This type of approach describes a network organization in which signaling is performed between network nodes and between associated KMS entities.

As shown in Fig. 5.8, the source node contacts the network controller with a request to obtain instructions for communicating with the destination node. In the case of MPLS, the controller responds but without providing details of the key which will be used. To obtain the key, the source node contacts its associated KMS. The KMS will communicate with its peer and negotiate the requested key material. Having received a response from the KMS, the RSVP signaling protocol is used to convey the keyID value written in the Session ID of the ERO subobject. Upon receiving the session ID, the destination node can contact its associated KMS and fetch the key material. The RESV message is returned to the source to signal a successful key establishment process which can trigger the establishment of a VPN tunnel [23].

Instead of using Path Computation Element Protocol (PCEP) MPLS communication between the nodes and a controller, an SDN OpenFlow protocol can be applied. SDN is categorized by relocation of the network node control logic to a centralized network controller. In this case, instructions from the higher layers of the network organization are given directly to the controller to establish a VPN tunnel between nodes. The controller dispatches requests via OpenFlow or NETCONF[2] messages to the network nodes, which then contact the associated KMS entities and obtain the key according to the procedure described previously [25, 26]. In addition, RSVP signaling does not need to be used since the controller can directly notify the destination node of the keyID obtained by the source node. The proposed model was later extended in the ETSI 015 specification [27].

We may observe that the described architecture involves the use of multiple signaling solutions. As shown in Table 5.1, the communication between nodes and the controller may include a certain QoS specification. Nevertheless, the signaling of specific QoS parameters is left for implementation by the dedicated signaling protocol used between network nodes. If the specification for communication is not dictated by the network controller, remote applications must harmonize the QoS specification before communicating with the KMS entity. Information about the route and state (congestion) of the network links which will be used to create the VPN tunnel can be obtained from the network controller. However, the specifications of QoS parameters, for example, permitted jitter when applications are served with keys or the negotiation of supported encryption algorithms, are not covered by this communication.

Table 5.1 Parameters exchanged between nodes and the controller [23]

Parameter	Description
KeyID	Key identifier used to synchronize the QKD key sessions
Key length	Length of the key used to provide a secure VPN tunnel
Destination	Destination node identifier
Source	Source node identifier
Encryption layer	TCP/IP layer in which encryption is performed
Encryption algorithm	Type of applied encryption cipher
Refresh type	Key refresh specification (i.e., time, traffic, length)
Refresh value	Key refresh specification value

---

[2] The main difference between the OpenFlow and NETCONF approaches is that due to the transactional implementation of NETCONF, it is not necessary to wait for a response on the success of the executed command since any misconfiguration or error will result in a configuration revert.

5.3 Summary

This chapter discussed different approaches applied in QKD networks for signaling, synchronization, and signal information exchange. Some of these solutions are suitable for distributed networks and take advantage of already authenticated packets from post-processing traffic to ensure the secure exchange of signaling information and a minimum of additional key consumption. The Q3P protocol emphasizes early organization of key storage by implementing dedicated STORE and LOAD subprotocols for key organization. These solutions have potential application in KMS–KMS communications. Sufficient quantities of keys must be provided for the smooth service of requests in accordance with the defined QoS parameters. However, in a hierarchical network structure, existing configurations of OpenFlow or NETCONF protocols can exchange the required details for obtaining the key. Although the ETSI 004 [24] and ETSI 014 [28] specifications have defined APIs and paved the way for obtaining keys from an associated KMS, it is obvious that signaling information throughout the process may require the use of multiple network protocols. Dedicated vendor-specific solutions may be unavoidable, therefore further complicating the conversion processes for defined requests. A larger group of solutions increases the risk of not supporting all the required QoS specifications. Nevertheless, the collected and exchanged signal and the signaling information could be of enormous use in defining the routes for the smooth distribution of cryptographic keys.

References

1. Armitage, G. (2000). *Quality of service in IP networks* (Vol. 1). O'Reilly & Associates. ISBN 1-57870-189-9.
2. Medhi, D., & Ramasamy, K. (2017). *Network routing: algorithms, protocols, and architectures* (Vol. 51). ISBN 978-0-12-800737-2.
3. Pióro, M., & Medhi, D. (2004). *Routing, flow, and capacity design in communication and computer networks*. Elsevier.
4. Yang, Y., & Wang, J. (2008). Design guidelines for routing metrics in multihop wireless networks. In *IEEE INFOCOM 2008-The 27th Conference on Computer Communications* (pp. 1615–1623). IEEE.
5. Rass, S., & König, S. (2012). Turning quantum cryptography against itself: How to avoid indirect eavesdropping in quantum networks by passive and active adversaries. *International Journal on Advances in Systems and Measurements, 5*(1), 22–33.
6. Kollmitzer, C., & Pivk, M. (2010). *Applied quantum cryptography* (Vol. 797). Springer Science & Business Media. ISBN 978-3-642-0482931-9. https://doi.org/10.1007/978-3-642-04831-9.
7. Mehic, M., Maric, A., & Voznak, M. (2017). QSIP: A quantum key distribution signaling protocol. In *Communications in computer and information science* (Vol. 785, pp. 136–147). ISBN 978-3-319-69910-3. https://doi.org/10.1007/978-3-319-69911-0_11
8. Lee, S.-B., Ahn, G.-S., & Campbell, A. T. (2001). Improving UDP and TCP performance in mobile ad hoc networks with insignia. *IEEE Communications Magazine, 39*(6), 156–165.
9. Wu, K., & Harms, J. (2001). QoS support in mobile ad hoc networks. *Crossing Boundaries-the GSA Journal of University of Alberta, 1*(1), 92–106.

10. Davidson, J., & Peters, J. (2000). *Voice over IP fundamentals*. ISBN 1-57870-168-6.
11. Sisalem, D., Floroiu, J., Kuthan, J., Abend, U., & Schulzrinne, H. (2009). *SIP security* (1st ed.). Wiley. ISBN 978-0-4705-1636-2.
12. Maurhart, O., Pacher, C., Happe, A., Lor, T., Tamas, C., Poppe, A., & Peev, M. (2009). New release of an open source QKD software: Design and implementation of new algorithms, modularization and integration with IPSec. In *Qcrypt 2013*.
13. Ghernaouti-Hélie, S., & Sfaxi, M. A. (2007). Upgrading PPP security by quantum key distribution. *IFIP International Federation for Information Processing, 229*(May), 45–59. ISSN 15715736. https://doi.org/10.1007/978-0-387-49690-0_4
14. Simpson, W. A. (1994). *The Point-to-Point Protocol (PPP)*. RFC 1661. https://rfc-editor.org/rfc/rfc1661.txt
15. Schartner, P., & Rass, S. (2009). How to overcome the 'Trusted Node Model' in Quantum Cryptography. In *2th IEEE International Conference on Computational Science and Engineering, CSE 2009* (Vol. 3, pp. 259–262). ISBN 978-0-7695-3823-5. https://doi.org/10.1109/CSE.2009.171
16. Schartner, P., Rass, S., & Schaffer, M. (2012). Quantum key management. In *Applied cryptography and network security*. InTech.
17. Peev, M., Pacher, C., Alléaume, R., Barreiro, C., Bouda, J., Boxleitner, W., Debuisschert, T., Diamanti, E., Dianati, M., Dynes, J. F., Fasel, S., Fossier, S., Fürst, M., Gautier, J.-D., Gay, O., Gisin, N., Grangier, P., Happe, A., Hasani, Y.,...& Zeilinger, A. (2009). The SECOQC quantum key distribution network in Vienna. *New Journal of Physics, 11*(7), 075001. ISSN 1367-2630. https://doi.org/10.1088/1367-2630/11/7/075001
18. Dianati, M., Alléaume, R., Gagnaire, M., & Shen, X. (Sherman). (2008). Architecture and protocols of the future European quantum key distribution network. *Security and Communication Networks, 1*(1), 57–74. ISSN 19390114. https://doi.org/10.1002/sec.13
19. Poppe, A., Peev, M., & Maurhart, O. (2008). Outline of the SECOQC quantum-key-distribution network in vienna. *Journal Of Quantum Information, 6*(2), 10. ISSN 0219-7499. https://doi.org/10.1142/S0219749908003529
20. Tanizawa, Y., Takahashi, R., Sato, H., Dixon, A. R., & Kawamura, S. (2016). A secure communication network infrastructure based on quantum key distribution technology. *IEICE Transactions on Communications, E99B*(5), 1054–1069. ISSN 17451345. https://doi.org/10.1587/transcom.2015AMP0006
21. Xianzhu, C., Yongmei, S., & Yuefeng, J. (2011). A QoS-supported scheme for quantum key distribution. In *2011 International Conference on Advanced Intelligence and Awareness Internet (AIAI 2011)* (pp. 220–224). IET. ISBN 978-1-84919-471-6. https://doi.org/10.1049/cp.2011.1461
22. Aguado, A., Lopez, V., Lopez, D., & Martin, V. (2017). Experimental validation of an end-to-end QKD encryption service in MPLS environments. In *Qcrypt 2017* (pp. 3–5). Cambridge.
23. Aguado, A., Lopez, V., Martinez-Mateo, J., Peev, M., Lopez, D., & Martin, V. (2018). Virtual network function deployment and service automation to provide end-to-end quantum encryption. *Journal of Optical Communications and Networking, 10*(4), 421. ISSN 1943-0620. https://doi.org/10.1364/JOCN.10.000421
24. European Telecommunications Standards Institute. (2020). *Quantum Key Distribution (QKD); Application Interface*. https://www.etsi.org/deliver/etsi_gs/QKD/001_099/004/02.01.01_60/gs_qkd004v020101p.pdf. ETSI GS QKD 004, v.2.1.1; Retrieved September 29, 2021 https://www.etsi.org/deliver/etsi_gs/QKD/001_099/004/02.01.01_60/gs_qkd004v020101p.pdf
25. Aguado, A., Lopez, V., Brito, J. P., Pastor, A., Lopez, D. R., & Martin, V. (2020). Enabling quantum key distribution networks via software-defined networking. In *2020 24th International Conference on Optical Network Design and Modeling, ONDM 2020*. https://doi.org/10.23919/ONDM48393.2020.9133024
26. Aguado, A., Lopez, V., Lopez, D., Peev, M., Poppe, A., Pastor, A., Folgueira, J., & Martin, V. (2019). The engineering of software-defined quantum key distribution networks. *IEEE Communications Magazine, 57*(7), 20–26. ISSN 15581896. https://doi.org/10.1109/MCOM.2019.1800763

27. European Telecommunications Standards Institute. (2019). *Quantum key distribution (QKD); Control interface for software defined networks*. https://www.etsi.org/deliver/etsi_gs/QKD/001_099/015/01.01.01_60/gs_qkd015v010101p.pdf. ETSI GS QKD 015, v.1.1.1; Retrieved September 29, 2021 https://www.etsi.org/deliver/etsi_gs/QKD/001_099/015/01.01.01_60/gs_qkd015v010101p.pdf
28. European Telecommunications Standards Institute. (2019). *Quantum key distribution (QKD); protocol and data format of rest-based key delivery API*. https://www.etsi.org/deliver/etsi_gs/QKD/001_099/014/01.01.01_60/gs_qkd014v010101p.pdf. ETSI GS QKD 014, v.1.1.1; Retrieved September 29, 2021 https://www.etsi.org/deliver/etsi_gs/QKD/001_099/014/01.01.01_60/gs_qkd014v010101p.pdf

Chapter 6
Quality of Service Routing in Quantum Key Distribution Networks

The basis of all modern networks is device which is able to manage and control network traffic. The fundamental network device is a router that connects at least two networks and has traffic directing functions. From a logical point of view, a router is the intersection of links forming a network. Its task is to accept incoming packets, analyze destination addresses, consult with its internal information how to process packets, and apply specified actions (discard it or eventually direct it to outgoing link).[1]

The distance between two geographic locations in road traffic can be measured according to various metrics, for example, distance in kilometers, number of administrative districts between locations, number of stops along a route, or the time spent on the road. Similarly, the distance between two routers in IP networks can be measured according to the number of intersections (hops) that a packet needs to pass to reach the destination, Euclidean distance, waiting time in queues in network devices, or other parameters. A generic term for the cost of traversing the network link is *routing metric*, which describes link statuses with precise numerical values. In the first ARPANET routing solutions, the metric was based on queue length since links with shorter queues were less congested and therefore preferred. However, these solutions, which were based on the distance vector framework, were soon abandoned because queue occupancy changes dynamically. Because of problems with routing loops, ARPANET considered other metric proposals in the late 1980s and switched to link state protocols. The protocols calculated the differences between time stamps when packets were received and when they were processed according to the overall processing delay of the router. They soon proved to be good indicators of the current link state but not good estimators of future values (especially for heavy load traffic). A metric which ignored the oscillations between

---

[1] In IP networks, the logical connection of a link is referred to as an interface which has its own determinants, such as an IP address, an MTU which defines the size that can be sent over the link, the statistics of the exchanged traffic, and other parameters.

© Springer Nature Switzerland AG 2022
M. Mehic et al., *Quantum Key Distribution Networks*,
https://doi.org/10.1007/978-3-031-06608-5_6

time windows using an M/M/1 queue formula to estimate the link state based on current load and variations in delay was applied instead [1].

A common factor in all these approaches was that routing was based on the use of a single metric to describe the link state and to calculate the shortest route. By contrast, QoS routing is the process of finding a feasible route (if one exists) so that multiple QoS metrics of interest remain within specific bounds.

6.1   Routing in General

Routing aspects can be categorized according to routing architectures, protocols and algorithms. The term *routing protocol* is often used as a comprehensive term for the overall solution, but more accurately, it denotes the actions and mechanisms which collect information about the network state. It defines how to structure routing packets, how to organize routing entries and databases, the period for which routes are valid, how to manage and update these routes, and when to remove them. The method of exchanging information and the method of storing and organizing the routing records depends on numerous factors, such as network dimensions, robustness, dynamics (which need not be reflected only in the node's mobility), scalability, computational and energy capacity, and others. Some routing protocols include a periodic exchange of routes to the first neighboring nodes. Some include trigger packet updates to exchange information about an occurred event while others include periodic link-state announcements based on network flooding. In this chapter, we consider the two most popular routing protocol classes: distance vector and link-state.

6.1.1   Routing Algorithms

When routing tables are filled with accurate information about the state of the network, the routing computation algorithm is used to calculate the route. The route can be determined manually, and in that case, routing is referred to as *static*. However, in most cases, it is desirable to use a routing algorithm which dynamically calculates the route based on defined traffic requirements, aims and collected information about the availability of network links.

Of the specific types of routing algorithms, Dijkstra's algorithm and the Bellman-Ford (BF) algorithm [2] are the main ones applied in practice. Both classify as user-oriented algorithms, i.e., they attempt to determine the shortest path from a location of the user's choice to other destinations. However, there is a network-oriented approach which attempts to achieve fairness between all network users and evenly maximize the use of every available network link. Dijkstra and BF are shortest-paths algorithms which find the shortest route from source to destination but account for concurrent traffic only to the extent of the impact on link "cost", which is the aforementioned metric to describe the link state. These link costs are furthermore assumed to be additive, meaning that the costs of two links add if the links are consecutively traversed by a packet. Therefore, "bandwidth" would not be suitable here since the overall path cost is determined by the minimum bandwidth over all segments, while "delay" would be an admissible cost since delays accumulate as the packet takes more and more hops. Dijkstra's algorithm requires that all link costs are non-negative, but the graph may contain cycles. BF, however, is able to find the shortest route even when negative costs for links are present [2], provided no negative cycles exist in the graph.

Here, it is important to separate the routing protocol from the routing algorithm. Although in practice the solutions that bind a certain protocol to an algorithm are most often present, solutions that do not have to follow this relation are also available. For example, some routing protocols are based on link-state classes that do not use Dijkstra's algorithm to find the shortest path. Examples are MPLS or GMPLS network environments which rely on Open Shortest Path First (OSPF) routing protocols to distribute information but do not require the use of the Dijkstra algorithm. There are also routing protocols which do not require information propagation at all (more details in Sect. 6.5).

6.1.2 Routing Architecture

Routing architecture defines the structures which organize the routing entities. More precisely, it defines the logical organization by considering the locations where the decision along the route occurs. Based on the location where the route calculation is performed, routing protocols can be divided into three broad categories: source routing, hierarchical routing, and distributed routing.

When source or explicit routing is applied, each node maintains an overall view of the network state. The source node calculates the route, which is written in the data packet header: this approach is often defined as *route pinning*. Intermediate nodes relay packets according to the information obtained from the header of received packet, but it also has the option to change the pinned route with the new route. The performance of source routing algorithms relies on the accuracy of precise state information.

In hierarchical routing, nodes are organized into groups which are recursively merged into virtual higher-layers, thereby creating a multilayer hierarchy. In each layer of the hierarchy, independent routing algorithms may be used. Hierarchical organization can define the node grouping into subdomain clusters to achieve scalability or robustness. The representative node from the group then participates in the routing decisions in the higher layer. Several examples which mainly follow the SDN approach demonstrate the use of hierarchical QKD network organization [3–6].

When distributed routing is applied, computation of the route occurs at each network node. The exchange of necessary routing information may be periodic in the case of a proactive (table-driven) protocol or only when a route is requested in the case of a reactive (on-demand) protocol. Proactive routing protocols mainly use fixed time periods which define how often to perform route updates. In overlay networks, reactive routing is more efficient and stable than proactive routing [7, 8].

6.2 Routing Requirements in QKD Networks

Routing in QKD networks is addressed on multiple layers. Consider the example network topology shown in Fig. 6.1a. Optical fiber connectivity is indicated with black lines, while the source and destination nodes and logical connection are marked in red. To establish a connection between these remote nodes, the routing protocol (e.g., OSPF, Routing Information Protocol (RIP), Intermediate System to Intermediate System (IS-IS)) must find the route over the existing IP network, as shown in Fig. 6.1b. This routing protocol can be used to define the route over which the encrypted traffic will be directed by applying one of the above-mentioned approaches (IPsec, Media Access Control security (MACsec), Transport Layer Security (TLS), etc.). It can be said that this type of routing protocol is in charge of defining the route for delivering secured traffic, i.e., using the previously distributed key material.

However, the route for the distribution of keys between remote nodes is not necessarily required to follow the route calculated by the IP routing protocol. The amount of key material determines the availability of the QKD link, leading to conclusion that the route defined for the consumption of keys might not be suitable for key establishment. An additional routing mechanism must therefore be used to determine the optimal route for key delivery, as shown in Fig. 6.1c. Additionally, when QoS routing is considered, the number of keys in QKD storage and additional parameters which characterize Quality of Service (QoS) connections (such as those briefly described in Sect. 2.2) should be considered. In this chapter, we focus on solutions which seek a route for the distribution of QKD keys between remote network nodes.

A routing protocol well-suited for operation in a dynamic QKD network should fulfill the main design objectives listed by priority as follows:

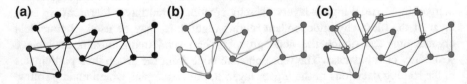

Fig. 6.1 Connectivity in a QKD network: (**a**) fiber topology and logical end-to-end connectivity; (**b**) data connection defined by the IP routing protocol (**c**) route for establishing the QKD cryptographic key defined by a quantum routing protocol

- By modifying the contents of routing packets, an attacker can redirect traffic to the node under his control [9, 10]. Given that the primary objective of QKD is to provide ITS communication, the routing packet therefore needs to be encrypted or at least authenticated [11]. This entails that the number of routing packets be minimized to preserve key material.
- The nature of QKD means that an eavesdropper is not able to gain information about the key which is transported via a link, and in the best case, a denial of service can be performed on the QKD nodes. To prevent such an attack, it is necessary to minimize the number of nodes which possess knowledge of the applied route, and thereby the number of routing packets which are broadcast are minimized [10].
- If the routing metric considers the key generation rate, the states of both QKD channels (public and quantum) should be considered since these channels are mutually dependent (see Chap. 4 for more details). Evaluation of the key generation rate only on quantum channel settings might be inaccurate given that communication over a public channel can significantly affect the duration of the key generation process.
- Since the number of nodes may vary for various realizations of the QKD network, the routing protocol should be scalable to different network sizes.
- Link interruptions are common in a QKD network due to its dynamic low key charging rate. Hence, the routing protocol should be robust enough to find an adequate replacement route in these situations.

The requirements listed above are related to the well-known Confidentiality-Integrity-Availability (CIA) scope of information security: confidentiality to maintain information secrecy, integrity to restrict changes to information, and authenticity to confirm the identity of users.

In a QKD network based on trusted-relays, each node implements routing and forwarding functionalities [12, 13]. The organization of these networks was therefore considered distributed without any purposefully hierarchical parent nodes. Each node implements routing functions and individually calculates the optimal key delivery route according to the information collected from neighboring nodes. Some of the solutions developed for this type of network organization are discussed in Sect. 6.5.

Nevertheless, the tendency of QKD technology to converge with existing telecommunications ISP networks is increasingly advocating a hierarchical approach. The KMS can be seen as a superior hierarchical entity which, in collaboration with the routing entity, determines the best route for key distribution. Under the EU H2020 OPENQKD project, the structure of a QKD network node [14] as shown in Fig. 6.2 has been proposed.

The KMS is the first contact-point for processing requests collected from key-seeking applications. It is linked to a QKD control entity which has the ability to control and monitor the QKD system by performing common operations such as power-on, power-off, reboot, restart, QBER/temperature measurement, and more. The KMS is aware of the state of QKD systems which establish QKD links with

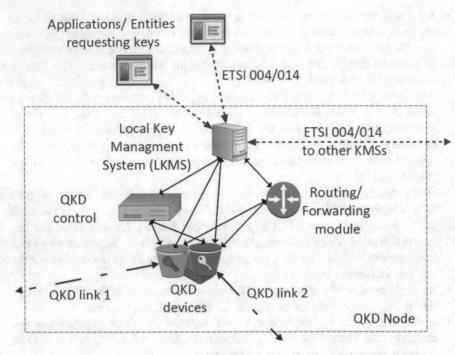

Fig. 6.2 Logical structure of components forming the QKD node

neighboring nodes and is able to communicate with other KMS entities to exchange management information. This type of communication can follow ETSI 004/014 or other specifications. In addition, the KMS of the node (often denoted *local KMS*) is able to communicate with a dedicated routing entity which attempts to calculate the optimal route for key distribution based on the information collected concerning the QKD link states and application requirements. It is important to highlight that a routing/forwarding entity should be physically independently installed of the KMS entity, which would provide a high degree of flexibility. Separation of the routing component allows the network administrator to replace or upgrade the routing logic when required without disrupting communications between other entities within the network node.

The arrangement described above can be described as gradually layered, given that in some locations only QKD devices can be implemented without additional network functionalities. Only some locations can have QKD systems which generate keys, but a single dedicated KMS which manages all other systems can be located in a different physical location. Provided the components are located within a secure perimeter, i.e., as long as communication between them can be performed reliably, it is possible to organize the node in such a manner.

If a network of such defined nodes is considered, it could be formed in a hierarchically organized manner. A dedicated KMS entity can be implemented to manage a network domain consisting of multiple network nodes. Figure 6.3 shows that one (orange) master KMS entity is implemented in domain 1 to communicate

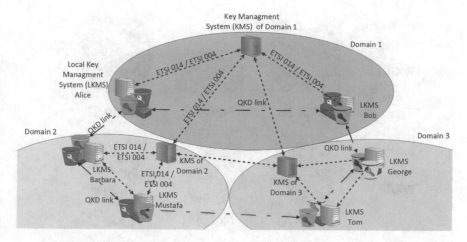

Fig. 6.3 KMS network organization

with local KMS entities but also with remote KMS entities in other domains. Organization in this manner refers to a hierarchical arrangement of the KMS entities. A single Local Key Manager System (LKMS) can manage multiple QKD systems.

In addition to hierarchical organization, an approach exists in which a KMS can be organized in a distributed manner, as with the case of domains 2 and 3 shown in Fig. 6.3. All KMS entities are interconnected, forming a distributed network, with the proviso that only a selected KMS entity may communicate with the KMS entities of other domains. This method of forming a network greatly affects how routing is performed. Routing protocols can therefore be based on a distributed or hierarchical organization or a combination of both.

QoS parameters which are of interest in a QKD network can be grouped into either additive or non-additive parameters. Namely, the basic QoS parameters parameters defined in Sect. 2.2 include the available amount of keys in key storage, delay and jitter in delivering the keys, and other specific parameters such as the type/vendor of device which generates keys, etc. In this context, it can be observed that the number of keys in storage is the primary parameter which dictates the existence of a route. An analogy would be the bandwidth in conventional IP networks.

Delay and jitter are classed as additive metrics because the total values of the route are obtained by summing the values of every link along the route. Bandwidth, however, is a non-additive concave metric limited by the bandwidth of lowest link in the route. From a computational point of view, the shortest routing path can be applied for delay and jitter, while the bandwidth requirement metric falls under the widest path routing.[2] Specifically, to ensure that the routing protocol converges

---

[2] Often referred to as the *maximum available trunk routing* or *maximum residual capacity routing* because of its use in dynamic call routing in telephone networks in the early 1980s.

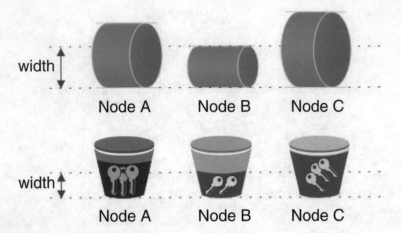

Fig. 6.4 Visual representation of the width of a path. Different bucket colors denote different keys along the path

toward the destination node[3] and satisfies the requirements for minimizing key consumption, widest-shortest path routing must be applied.

The similarity between bandwidth and the number of keys in the QKD repositories is shown in Fig. 6.4. Referring to Sect. 3.1, we can link the number of keys in QKD storage with the number of available slots in buffers on nodes in the IntServ architecture [15]. However, there is a difference, as these buffer calculations are dictated by message signaling instead of the exchange of information through routing packets.

Regarding dedicated QoS parameters described in Sect. 2.2, for example, epsilon security, the type of device which generated the key, or the path through which the key was distributed, these parameters can be taken into account when obtaining keys from key repositories. Namely, if some of these requirements are set, they will result in additional key filtering, and the feedback will indicate the quantities of filtered keys or the delay/jitter which occurred as a result of applying these filters.

6.3   Addressing in QKD Networks

The main task in implementing routing is the allocation of addresses to network nodes. The routing protocol should have knowledge about the node identifiers to be

---

[3] For illustrative purposes, consider routing in distributed networks if it were based on routing only considering the amount of available bandwidth. Routing such as this might distract the packet from the destination if the links leading to the destination have poor bandwidth. Consequently the packet may never reach the destination.

able to form routes to establish connections. Consider, for example, the three-node topology shown in Fig. 6.5.

Nodes can have different identifiers and potentially be allocated an address based on arbitrarily assigned names (Alice, Bob, Barbara, and others). However, in practice, QKD nodes implement at least two interfaces: public channels and quantum channels. Both channels can be practically realized through the same medium, but logically they are separate channels (Sect. 2.4).

When two nodes are directly connected (forming a single point-to-point link), a routing protocol is not required. However, if it is necessary to distribute keys between multiple remote nodes which are more than two hops away, a routing protocol is indispensable. In trusted-relay QKD networks, routing via quantum channels is not an option. The QKD system identifiers of quantum channels are therefore insufficient. The keys are distributed via public channels created using standard TCP/IP interfaces, while identifiers of public channels may be considered.

In practice, a node can implement multiple network interface cards which provide multiple TCP/IP interfaces. It is also possible to use the same physical TCP/IP interface to create multiple connections. The single interface can be used to implement multiple public channels via virtual TUN/TAP TCP/IP interfaces forming dedicated VPN tunnels. TCP/IP interface identifiers are those which uniquely specify the node to which the key must be delivered and are used generally to implement the allocation of addresses to QKD networks. Since the implementation of QKD networks is not expected to exceed metropolitan scales, it is sufficient to use existing IPv4 network addresses [16]. In addition, the QKD network is usually implemented as a private network, which allows greater freedom to manage address spaces.

6.4 Routing Protocols

One major classification forms two groups of routing protocols: static and dynamic (equivalently non-adaptive and adaptive). In static (non-adaptive) routing, protocol policies are established at the beginning (OS configuration) and do not react to network changes (topology, weights, etc.). In dynamic (adaptive) routing, routing tables are instead calculated according to the information collected on the network's

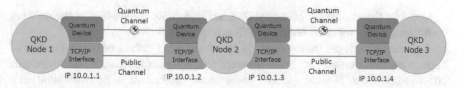

Fig. 6.5 Example of QKD device connectivity with the specified IP addresses of the TCP/IP network interfaces. They uniquely determine the link and QKD system which establishes the key and can be used to uniquely identify the node

topology, the cost of paths and the state of the network's components (these criteria change over time). The more interesting and most widely used protocols are dynamic and can be either centralized or distributed. In centralized routing, each node receives all information from the network and distributes the routing tables to the nodes after calculation. In distributed routing, the nodes exchange information to calculate the routing tables collaboratively.

In this chapter, we list the basic features of dynamic protocols with an analysis of their application in QKD networks.

6.4.1 Distance Vector Routing Protocols

The Distance Vector (DV) protocols are the oldest routing solutions. They are often referred to as BF protocols because its most common practical application is with the BF algorithm [17].

DV protocols are based on vector propagation (hence the name) of the distance cost values. Each router maintains a local database with a minimum distance between itself and all known destinations. At regular time intervals, it broadcasts a data-structure and the distance-vector, which is a set of address-distance pairs (announcements), to its neighboring router. By default, the distance denotes the single-value cost of reaching a remote network node, expressed as a number of hops or with other general criteria which may take into account the bit-rate, load or connection reliability. When a network router receives broadcast data, it can use the BF algorithm to create a routing table which associates the cost to reach the node and the first hop of the calculated route to each known destination. A router recalculates its routing tables under any of the following circumstances:

- The network interface state changes to up or down,
- The router receives an announcement from a neighboring router for an unknown destination,
- The router receives an announcement from a nearby router for a known destination but with a lower cost than the stored one,
- The router receives an announcement from a neighboring router for a destination that the same router had previously announced with a higher cost,
- The Time to Live (TTL) for a destination in the table expires.

Consider the example topology shown in Fig. 6.6. Routing tables in the network convergence state for each node u are shown in Fig. 6.7.

As noted above, the cost values do not necessarily reflect only the number of hops to reach a destination. With frequent and significant changes, however, side effects may occur. Several specific phenomena are known to be associated with distributed information propagation. Temporary occurrences of temporary routing loops[4] or "bouncing effects" are possible at moments when the routing protocol has

Fig. 6.6 Example network topology

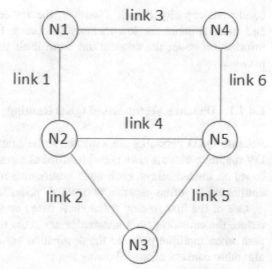

Fig. 6.7 Associated distance-vector tables on nodes

From N1 to	Link	Cost
N1	local	0
N2	1	1
N4	3	1
N3	1	2
N5	1	2

From N2 to	Link	Cost
N2	local	0
N1	1	1
N4	1	2
N3	2	1
N5	4	1

From N3 to	Link	Cost
N3	local	0
N2	2	1
N1	2	2
N5	5	1
N4	5	2

From N4 to	Link	Cost
N4	local	0
N1	3	1
N2	3	2
N5	6	1
N3	6	2

From N5 to	Link	Cost
N5	local	0
N2	4	1
N1	4	2
N4	6	1
N3	5	1

not completely converged to the new stable state. Some nodes use outdated link cost values because new information has not yet been delivered to them [1].

DV protocols are more suitable for small networks without stringent performance requirements. Finding the best route requires more time, given that knowledge of the entire topology is not immediately available for the calculation, which in practice translates complexity into lower convergence speed. Updates are also sent periodically, not simply when a network event which affects the overall network

---

[4] A packet is bound to moving between two or more routers without obtaining an exit.

bandwidth capacity occurs. However, the router configuration is much simpler and more suitable for low-memory resources. Each router needs only to store information about the connections from itself to the other nodes, not the entire network.

6.4.1.1   Distance Vector-Based QKD Routing

Because QKD networks are expected to be limited to a metropolitan scale [13], DV routing protocols have been identified as a possible solution. In QKD networks based on trusted relays, each node implements routing functionality. All network entities are therefore generically denoted "nodes" in the following text.

One of the first routing solutions is based on stochastic path selection [18]. To reduce the eavesdropper's knowledge about the route, each node selects a random path when multiple paths to the destination are available. The stochastic routing algorithm consists of the following steps:

- Nodes collect information about the available neighbors and their link statuses (available number of keys).
- When a packet reaches a relay (intermediate) node on a source-destination route, the relay node calculates all possible routes to the destination using the shortest-path algorithm. Although the type of algorithm and method of distributing routing data is not explicitly stated, the proposed solution is compared with the well-known RIP distance-vector routing protocol.
- If several routes to the destination are available, a randomly selected link which has sufficient key material is selected to forward the packets. If not enough key material is available at any of the outgoing links, the forwarding algorithm is aborted.

The authors of [18] briefly described the results of simulations from which they claim that the use of stochastic routing may lead to increased packet loss rate.[5] Packets can be routed to nodes which have no available outbound links with a sufficient number of keys. Forwarding packets are then aborted, producing reduced routing performance to the detriment of the attacker's knowledge of active paths. Regardless, the stochastic routing approach is one of the first QKD routing solutions which attempts to balance the network load by evenly distributing traffic. It uses information only about the number of keys on the links to neighboring nodes as a criterion for determining the quality of the path. This type of approach is a heuristic method of determining the route, but it does not guarantee that it really exists. It may result in no route being available after filtering, which is dictated by the number of available keys in the key buffers.

Inspired by the similarity of QKD and ad hoc wireless networks (Sect. 2.5), a simple approach to simulating QKD networks was proposed in [19]. Since the QKD

---

[5] Packet loss rate is defined as the ratio of received and sent packets.

link may be considered available only when sufficient keys exist for encrypting the transmission of content over that link, the settings for point-to-point links can be used to reflect its availability. The authors of [19] suggested that when sufficient keys are available on the link, the propagation delay of the point-to-point link should be set to 2 ms, which is a common value for point-to-point links [20]. Otherwise, the link delay is set to 100 seconds, which in practice means that the link is unavailable but will re-grow key material over time (remember that QKD key exchange is intended to run endlessly and continuously). Once the key stores have been accordingly refilled, the link delay is returned to the initial value of 2 ms. This type of simulation approach has implications. Specifically, setting a delay link to larger values disables any practical communication via that link. In practice, it is not realistic because communication which does not require any consumption of keys (such as some of post-processing modules) may exist. This also implies that routing packets still see that link as available for traffic forwarding (the interface is active). Link performance is significantly impaired as a result of increased point-to-point delay, but the routing protocol is not instantly aware of it. The routing protocol can detect when a link is unavailable from the lack of "hello" routing packets exchanged over the link. The primary aim of these packets is to refresh existing routes in the routing tables and confirm the presence of adjacent nodes. The example described suggests that the validity of the routes should be based on the amount of available key material.

Another potential approach in simulating QKD links is to simply invoke TCP/IP interface shutdown actions instead of increasing link latency. Such an action would, however, lead to frequent flooding of the network with triggered routing updates that exchange information about the availability of network links. Additionally, such actions would prevent the generation of new keys which are based on a timer-based post-processing application (periodic buffer filling which requires knowledge of peer connections).

In implementing the simulation environment using the approach described above, research has further explored the possibility of applying routing protocols originally developed for ad hoc networks. Compared to the OSPFv2 link-state solution predominantly used in DARPA and SECOQC quantum networks [16, 21], the Destination-Sequenced Distance-Vector (DSDV) protocol has been shown to produce convincing results.

DSDV is the most widely applied DV protocol based on the BF algorithm. Each node implements a permanent routing table containing all the destinations which can be reached within the network, the address of the next hop, and the cost value defined with the total number of hops needed to reach the destination. By default, each node periodically broadcasts its routing table to its neighboring nodes every 15 seconds. Upon receiving the periodic update packet, the neighboring node will update its routing table. For each record, the number of hops increases by one. The update is then forwarded to neighboring nodes, and so on, until all network nodes have received a copy of the updated value. To manage routing loops, DSDV implements sequencing numbers which indicate the freshness of the information passed through the network. The higher the sequencing number, the fresher the information. When two updates are received, the update with the lower sequential number is abandoned. Although periodic updates are frequent, faster propagation of urgent information is required. This is achieved using triggered update packages which urgently notify the status of the link, such as the appearance of a new node, shutdown of the network interface, etc. Frequent changes in the network link status may result in the dominance of triggered packets. Settling timers can therefore be introduced to retain triggered packets for a brief period and converge the network to a stable state. An analysis of the resulting performance, demonstrates that even without specific QoS mechanisms, the DSDV protocol achieve noticeable performance improvements. Periodic exchange of complete routing tables every 15 seconds leads to a sufficiently up-to-date network state with small packet-loss values. It also reduces the number of routing packets which are encrypted for protection against abuse.

6.4.2   Link State Routing Protocols

The Link-State (LS) based routing is mostly combined with Dijkstra's algorithm, which requires topological information about the nodes forming the network and the links connecting these nodes to calculate the shortest path. Each link has its own cost and information whether the link is available, commonly referred to as the *link state*. A routing protocol based on an approach in which all network nodes possess information about the state of all links within the network is therefore referred to as a *link-state* routing protocol.

Unlike delivering vectors which contain calculated information about possible destinations in a distributed fashion, LS routing protocols are based on the concept of distributed maps. This means that all nodes maintain a regularly updated network topology map. With accurate information in the map, each node can calculate the route not only to a particular destination but to all nodes in the network.

The network map is built by routers using Link-State Advertisement (LSA) packets: each router broadcasts information about which nodes are adjacent to it. It performs selective flooding by retransmitting the packet to all interfaces except the interface over which the message is received. Each node contains an LS database to store the most recent LSA generated by each node (the databases converge to be identical on each node in the network). The LSA contains information such as source node identifier, link identifier, link cost, sequence number and age.

Links in the link state routing protocol are marked as directional, i.e., the cost values of link $x \rightarrow y$ might differ from the cost values of link $y \rightarrow x$. Sequence numbers in LSA are used to indicate the freshness of the LSA update, since obsolete LSA information might be delivered multiple times because of flooding mechanisms. Each time a node prepares an LSA for an specific link, it increases the sequence number counter and stamps the new value in the LSA packet. However, the sequential numbers count the modulo of the maximum value to fit into the range of the sequence field so that the counter resets to zero if an overflow occurs. To avoid confusion with LSA packets generated by nodes which have recently joined the network or nodes on which the routing protocol has been reset and do not have complete information about the network topology, the field *age* is used to indicate the freshness of the information. The node which generates the LSA packet writes the maximum value in the age field, which is decremented each time it is forwarded. If, however, it reaches a value of zero, it is considered obsolete and no longer forwarded.

To minimize flooding the network with a large number of LSA packets which describe each individual link, a node may form an Link-State Update (LSU) packet containing multiple LSAs received from multiple neighboring nodes. In addition to LSA packages, the LS protocol often implements two subprotocols which generate two other types of message: hello and resynchronization. The hello subprotocol is used when initializing a node to notify neighbors of its presence. It is also used to gather information about the rest of the network so that a topological map can be formed and the routing table recalculated. It is executed periodically to check the operation of links to neighboring nodes. Similarly, the resynchronization subprotocol is used after link or node failure. During failure and reactivation of a link, multiple values of the link state can be detected. A resynchronization subprotocol is therefore applied to collect the latest up-to-date state information that will result in a fresh LSA packet.

The LS database can be considered a complete representation of the network, and as such, it is enough for each router to calculate the routing table autonomously (in contrast to distributed approaches). Each LS database record contains the identifiers for the departure and arrival routers, the link identifier and the relative cost distance. Since all routes have the same network database, computation of the routes is coherent and without routing loops. The example topology and associated link-state table is shown in Fig. 6.8.

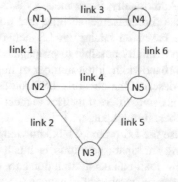

From	To	Link	Cost
N1	N2	1	1
N1	N4	3	1
N2	N1	1	1
N2	N3	2	1
N2	N5	4	1
N3	N2	2	1
N3	N5	5	1
N4	N1	3	1
N4	N5	6	1
N5	N2	4	1
N5	N3	5	1
N5	N4	6	1

Fig. 6.8 Example network topology and associated link-state table implemented at each networking node

The timing of the LSA is an important detail, because when to build the LSA packets must first be decided. They can be generated either at regular intervals or when a significant event occurs (a new adjacent node is detected, the cost of a link has changed, a link fails or is restored after a fault, etc.). When a router receives an LSA from a particular neighbor, it must apply a series of actions:

• If the router has never received an LSA from that specific neighbor or if the LSA is more recent (sequence number) than the one stored, it stores the LSA and retransmits using the flooding technique,
• If the LSA has the same sequence number as the one stored, it performs no action,
• If the LSA is older than the one stored, it uses the most recent one for flooding.

In this manner, the LS databases of all routers are kept aligned and consistent to ensure correct routing (the network is said to be *in convergence*).

LS protocols have gained great popularity over DV solutions for several reasons:

• *Rapid convergence without the formation of routing loops*—Distributed calculation means that DV protocols require calculation on each network node after a new network event. The nodes without fresh information contain outdated information which, if used, may lead to the formation of routing loops. Nodes must also wait for the calculations to be performed on remote network nodes and the propagation of new information to them. LS protocols are faster because they employ local calculations on each network node in parallel. With instant calculation, LS protocols are also more efficient, reducing the time window for the appearance of routing loops.
• *Ability to apply multiple routing metrics*—In previous examples, we have stated that routing protocols rely on distance (number of hops) as a primary estimate of path quality. Nevertheless, cost metrics can be defined to encompass various parameters which describe the quality of link state. If the metric were defined in this manner, only one value of the calculated value is gradually distributed in the case of a distributed DV calculation. Additionally, if the metric value changes abruptly and frequently, it can increase the convergence time and potentially form routing loops. LS protocols are based on having an LS network map. When link states are distributed, it is potentially possible to propagate multiple cost values which describe their states (bandwidth, jitter, number of hops, etc.). In possessing multiple metrics which describe the network, LS protocols (e.g., OSPFv2) can perform route calculations using different metrics to meet different QoS requirements (such as best delay, best throughput, etc.).
• *Support for multipath routing*—Because the LS protocol has knowledge of the entire network topology, implementing multipath solutions is much simpler. Specifically, some LS protocols allow load-balancing functions to optimize network use if multiple paths with the same number of distances (costs) to the destination are available.
• *Support for multiple external routes*—In addition to routing within the local domain, enabling packet routing to external domains is often necessary. No major differences exist between the LS and DV protocols when only one external

gateway to the outside world is available. However if multiple output gateways are available, DV protocols will include one entry per destination in the DV packet, whereas LS solutions write entries into the *gateway link state records* database. More precise metrics and easier computation in the LS approach leads to better performance.

6.4.3 QKD Routing Based on Link-States

Given the widespread use of link-state protocols in conventional IP networks, it has been applied with minor modifications during the development of the first QKD test bed network. Similarly to DV approaches, some solutions modify the metric which defines link cost and others parse the shortest path using conventional algorithms and subsequently filter the path according to QKD critical parameters, such as the number of keys in storage.

Path calculation in the DARPA QKD network was based on Dijkstra's algorithm, the cost metric being defined according to [22]:

$$m = \begin{cases} 100 + \frac{1000}{q-t}, & q > t; \\ \infty, & t \leq t, \end{cases} \tag{6.1}$$

where m is the link cost metric, q is the number of available blocks of keys (Qblocks) on the link in one LSA update interval, and t is the configurable threshold (default value 5) for a minimum number of Qblocks on the link for authentication in post-processing communications. The larger the number of keys in storage q, the lower the cost value denoting the link as more desirable for forwarding traffic. The metric m does not take into account other parameters such as link load or delay [23].

To refresh the records in link-state databases, periodic messages ROUT1_LSA are exchanged [22]. These messages contain the node IDs of the sending node and neighboring node and the corresponding 32-bit link metric. ROUT1_LSA messages are exchanged at every LSA update interval (one minute by default). Each node has an individual LSA timer which does not depend on other nodes in the network.

The SECOQC network proposed in [16] employed a modified OSPFv2 protocol. Its main difference is that it computes several short routes for each destination. To satisfy the requirements for load balancing, multiple routes were needed, and therefore, each node calculated as many routing tables as its active interface.

With routing tables calculated for each interface, each node implements an *extended routing table* merging all entries arranged according to cost in increasing order. The table's layout is identical to the common routing table, the difference being that the numbers of outgoing nodes for each destination are equal to the number of entries [16]. The node can therefore select several routes to the target node, but it needs to know the approximate load of each connection in the path. A threshold value L^{cr} is therefore introduced to identify the validity of the route. If the connection load is greater than L^{cr}, then the next best route is applied.

To maintain information about the approximate load on each link, a *load status database* table is calculated [11]. A low pass filter is used to determine the approximate output load of the interface i at discrete t, denoted $L_i(t)$:

$$L_i(t) = \left(1 - \frac{1}{w}\right) \cdot L_i(t-1) + \frac{1}{w} \cdot \ell_i(t), \tag{6.2}$$

where w is the filter constant and $\ell_i(t)$ is the instantaneous load of the outgoing link i, which is defined by the number of transmitted bits in the most recently occurring period. The SECOQC approach is thus based on the calculation that at least one interface will be below the threshold value L^{cr} if multiple interfaces have been implemented on the link. In addition, the larger the number of packets sent over the interface, the greater the use of QKD material; Eq. (6.2) seeks to realize connections equally through all available interfaces. This approach is a variant of the stochastic approach defined in [18], but, although it strives to realize even use on all interfaces, it does not provide QoS guarantees.

In [24], the link metric is based on the time required to implement the key relaying process. The protocol is based on propagating link state values using a dedicated "hello" subprotocol which is in charge of flooding the initial information about nodes and network links and their settings (link ID, key generation rate, amount of residual key material in key storage). The cost metric C to denote the state of the link $e_{i,j}$ from node i to node j is calculated as

$$C(e_{i,j}) = \begin{cases} \left\lfloor \frac{N-M}{G_{i,j}} \right\rfloor, & (N-M) > G_{i,j} \\ 1, & otherwise, \end{cases} \tag{6.3}$$

where M denotes the available amount of key material, $N = \max(n^i_{e_{i,j}}), i = 1, 2, \ldots$ denotes the maximum key storage capacity of links (edges e) implemented between QKD nodes, and $G_{i,j}$ is the generation rate of the link. The values N and M are expressed in the amount of key material (bits), while G is expressed in the amount of key material (bits) per unit time resulting in a metric which describes the maximum time to perform a key relaying operation. Since this link cost metric actually represents a time unit, it is an additive type, and the total path metric of the selected key relay path from source s to destination d can be calculated from

$$C'(p_{s,d}) = \sum C(e_{i,j}), \tag{6.4}$$

where $p_{s,d}$ is the path from source s to destination d, and the sum passes over all edges constituting this path.

Note that the key relay approach requires the consumption of the same number of keys through each engaged link, therefore it is necessary to select the shortest path from the source to destination. To avoid routing over a path which is not the shortest, even though it has the most available key material at the given moment, the final path metric is corrected with

$$C(p_{s,d}) = C'(p_{s,d}) + f(p_{s,d}), \tag{6.5}$$

where $f(p_{s,d})$ is defined by

$$f(p_{s,d}) = \begin{cases} (n/5) \cdot \overline{C(e)}, & n \leq 5; \\ \overline{C(e)}, & n > 5, \end{cases} \tag{6.6}$$

$\overline{C(e)}$ being the arithmetic mean of all link costs along the path $p_{s,d}$ and n denoting the lowest number of hops. This approach attempts to define a $widest - short$ route which analyzes the number of hops across the path and the number of keys on the links forming the path. The authors of [24] proposed using Dijkstra's algorithm to calculate the shortest route based on metrics defined with (6.6). Since the key generation rates are low, the use of multiple calculated routes is suggested. For the purpose of maintaining information about QKD links, a modified LSA package is used. The LSA package is in charge of distributing information about the number of stored keys in storage and the key generation rate of the keys which are active. If the intensity of key consumption via the link and key generation are similar, advertising LSA packages is unnecessary since new traffic cannot be served via links which already have a maximum load. Similarly to previously described solutions, this approach is based on the use of key material as a basic determinant for route calculation.

Link-state routing has also been applied in several similar approaches which consider the amount of available key material as the primary link metric [25].

6.5   Greedy Perimeter Stateless Routing for QKD Networks

As described in Sect. 2.5, the similarities of QKD with ad hoc networks inspired the creation of Greedy Perimeter Stateless Routing Protocol for QKD Networks (GPSRQ), which is proposed in [26]. The main motivation for designing GPSRQ was to minimize the number of routing packets with regard to the requirement for minimizing key material consumption. GPSRQ is based on the Greedy Perimeter Stateless Routing in Wireless Networks (GPSR) routing protocol [27], with several significant changes.

As shown in Fig. 6.9, the key material storage of link k between nodes a and b can be represented by the following parameters:

- Time measurement interval t, measured in seconds,
- Mean key generation rate r_k, measured in bits per seconds and used to indicate the charging rate for storage,
- The key material storage depth $M_{max,k}$, used to indicate the storage capacity,

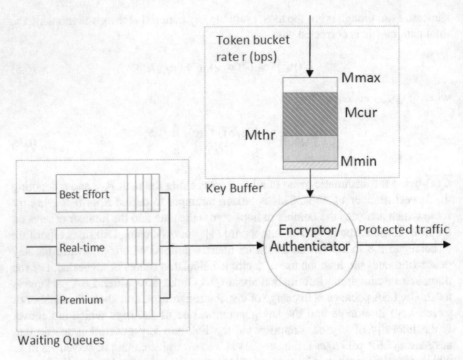

Fig. 6.9 Token bucket FQKD model

- The current value $M_{cur,k}(t)$, representing the amount of key material in storage at the time of measurement t, such that $M_{cur,k}(t) \leq M_{max,k}$.
- The threshold value $M_{thr,a,b}(t)$, or simply $M_{thr,k}(t)$,
- The minimum amount of pre-shared key material, denoted $M_{min,k}$.

The amount of key material in storage at the measurement time t can be bounded using the expression (6.7), while the average operational rate can be calculated using Eq. (6.8).

$$D_k(t) \leq r_k \cdot t + M_{cur,k}(t) - M_{min,k}(t) \tag{6.7}$$

$$A_k(t) = \frac{D_k(t)}{t} = r_k + \frac{M_{cur,k}(t) - M_{min,k}(t)}{t}. \tag{6.8}$$

The FQKD model described in Sect. 3.4 proposed the use of threshold value M_{thr}. This parameter is applied to increase the stability of QKD links, such that $M_{thr,k}(t) \leq M_{max,k}$. The meaning of this parameter is best explained by considering the simple topology illustrated in Fig. 6.10, where node a needs to establish a connection with remote node e. Suppose the routing protocol uses only information about the link-state with its direct neighbors. Then, assuming that every network

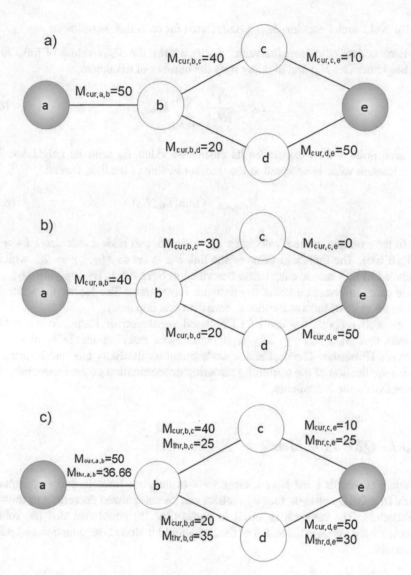

Fig. 6.10 Simple topology illustrating the calculation of M_{thr}: (**a**) traffic is routed along the path a-b-c-e; (**b**) traffic is routed along the path a-b-c, regardless of key material depletion at link c-e; (**c**) calculated M_{thr} values are indicated next to M_{thr} values

link has the same performance as a public channel, we consider only the status of key material in storage, which are marked next to the links (Fig. 6.10a). Upon receipt of the packet from node a, the routing protocol on node b selects path b-c toward destination e since the link b-d has poorer performance. However, since node b does not consider the link-state more than one hop away, traffic may become stuck on the link between nodes b and c (Fig. 6.10b).

To avoid such behavior, M_{thr} is calculated for each link as follows:

- Each node a calculates the value L_a by adding the M_{cur} values of links to its neighbors (set N_a) and dividing it by the number of neighbors,

$$L_a = \frac{1}{|N_a|} \cdot \sum_{j \in N_a} M_{cur,a,j} \qquad (6.9)$$

- Each node a then exchanges its calculated value L_a with its neighbors. The minimum value is accepted as the threshold value of the link, that is:

$$M_{thr,a,b} = \min\{L_a, L_b\} \qquad (6.10)$$

In the example, node b calculates $L_b = 36.66$, and node c calculates $L_c = 25$ (Fig. 6.10c). The threshold value of the link b-c is set to $M_{thr,a,b} = 25$, which is included in link metric calculation described in Sect. 6.5.1. By applying M_{thr}, the node gains information about the network links-states. The higher the value, the better the state of links at locations more than one hop away.

We also propose the use of ECN-based regulation of traffic. When a node detects that $M_{cur,k}(t) \leq M_{thr,k}$, it commences marking the ECN bits in the packet's IP header. The destination node monitors the ECN bits and informs the user's application of the upcoming rerouting or termination of the connection if no alternative route is available.

6.5.1   QKD Link Metric

Routing protocols must have a clear view of network links to define an optimal path. In a QKD network, routing metrics can be categorized according to multiple parameters. The research reported in paper [26, 28] concluded that the routing metric which assesses the status of the link should include both quantum and public channels.

6.5.1.1   Quantum Channel Status Metric

The amount of key material in key storage is the main factor contributing to the availability of a link. Equation (6.12) expresses the state of the quantum channel between nodes s and i, where $Q_{frac,s,i}$ is the ratio of the squared amount of key material at the time of measurement ($M_{cur,s,i}^2$) multiplied by the threshold value ($M_{thr,s,i}$) and the cubed capacity of the key storage ($M_{max,s,i}^3$) defined by Eq. (6.11). $Q_{frac,s,i}$ falls in the range [0,1] and denotes the available amount of key material on direct links in relation to the amount of key material of links which are further

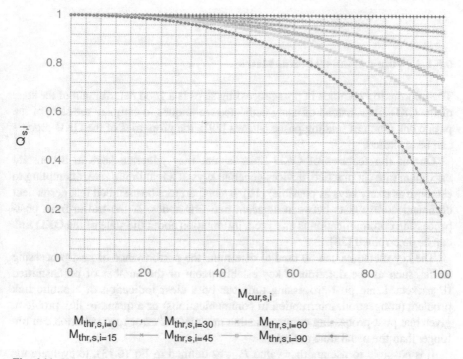

Fig. 6.11 Values of $Q_{s,i}$ of the QKD links between nodes s and i for different values of M_{thr}; $M_{max,s,i} = 100$

away, as those links are unreachable if direct links are unavailable due to lack of key material:

$$Q_{frac,s,i} = \frac{M_{cur,s,i}^2 \cdot M_{thr,s,i}}{M_{max,s,i}^3}, \qquad (6.11)$$

$$Q_{s,i} = 1 - \frac{Q_{frac,s,i}}{e^{(1-Q_{frac,s,i})}}. \qquad (6.12)$$

The value $Q_{s,i}$ is the utility in terms of the available key material of the link. Since no reservation mechanisms are in effect, the key material can be consumed rapidly if a large number of key use requests occur (from the same or different applications). $Q_{s,i}$ is therefore exponentially scaled to warn of the upcoming critical situation of low key material amount. If a small amount of key is available, a rapid response from the routing protocol is required to find an alternative path which is signaled through the exponential of $Q_{m,s,i}$.

The value of $Q_{s,i}$ is limited to the range [0,1] (Fig. 6.11), a lower value indicating a better quantum channel state. In the example in Fig. 6.10c), the routing protocol should therefore favor the link b-d since $Q_{b,d} < Q_{b,c}$.

6.5.1.2   Public Channel Status Metric

The information obtained in post-processing steps is a good determinant of the state of the QKD link. Instead of using dedicated techniques to estimate the state of the public link (such as sending probe packets [29]), the properties of the QKD process can be exploited.

One of the features of QKD links is that they generate keys at a constant maximum rate (in practice, devices generate keys at a maximum rate, attempting to establish as many keys as possible) [16]. It can therefore be assumed that permanent communication exists between adjacent network nodes. If the traffic from post-processing communication is analyzed, information about the state of the QKD link can be determined [28].

The key metadata can be used to determine the performance of post-processing traffic, such as the duration of key establishment or the number of retransmitted IP packets. Long post-processing time can be a clear indication of a public link problem (congestion, interruption in communications) or a quantum link problem, given that post-processing communication modules for error reconciliation can run longer than the usual time period.

It is possible to use another value $P_{s,i}$, as defined in Eq. (6.13), to evaluate the state of the public channel between nodes s and i: Let $T_{last,s,i}$ denote the time spent on the establishment of the key at the time of measurement, and let $T_{maximal,s,i}$ be the maximum time which can be tolerated for the establishment of the key. The value $T_{maximal,s,i}$ is set as double the value of the average duration of the key material establishment process in the long term, denoted $T_{average}$ in Eq. (6.14). Then, we compute

$$P_{s,i} = \frac{T_{last,s,i} + \Delta t}{T_{maximal,s,i}}, \tag{6.13}$$

with

$$T_{maximal,s,i} = 2 \cdot T_{average}. \tag{6.14}$$

The Δt parameter is used to describe the differences in measurement time and the time when the most recently recorded value $T_{last,s,i}$ was taken. The value $T_{average}$ depends on the type of quantum and network devices, the QKD post-processing application and the performance of the public channel. Low values of $P_{s,i}$ indicate better public channel states. Values greater than 1 indicate that the link has a problem with the establishment of new key material.

6.5.1.3   Overall QKD Link Status Metric

After considering the metrics for individual channels, a unique metric which describes the state of QKD links was proposed in [26]. Since the individual channel metrics do not consider the type of cryptographic algorithm for which the established key will be used, the parameter $\alpha \in [0, 1]$ was introduced, with the following use. Consider the link between nodes s and i with a small value of $Q_{s,i}$. This link may be suitable for network flow encrypted using algorithms which do not require large amount of key material (e.g., an AES cipher with large rekeying values). However, due to a lack of key material, this link is not acceptable for encrypting traffic flow using an OTP cipher. Hence, the value of α can be adapted to account for the "key demand" of different algorithms, leading to a modified metric:

$$R_{s,i} = \alpha \cdot Q_{s,i} + (1 - \alpha) \cdot P_{s,i}. \tag{6.15}$$

The joint equation (6.15) combines the utility functions of the quantum and public channels. Low output values indicate a better link state.

6.5.1.4   GPSRQ Routing Protocol

With defined QKD link metrics, the GPSRQ routing protocol can calculate the network routes. The fundamental assumption is that all nodes know the geographic locations of other nodes. While this may be a rigorous assumption for mobile QKD nodes, most nodes in a QKD network are static (due to the sensitivity of quantum equipment). Assuming that nodes are static, then network flooding to determine the position of network nodes can be avoided. Instead, we can use a location registry that is updated periodically if mobile devices are present on the network [30–34]. The use of post-processing traffic for the transmission of signaling geographic information has also been suggested, as some of the post-processing packages are already authenticated.

Starting with this fundamental assumption that the nodes are static, we can take advantage of the distance-dependent location update approach: "*the greater the distance separating two nodes, the slower they appear to be moving with respect to each other*" [35].

GPSRQ is designed for distributed networks which have no hierarchical parent nodes. Additionally, to avoid attacks where an attacker redirects traffic to the node under his control [10], GPSRQ does not exchange routing tables. An attacker is unable to intercept routing packets and does not know which network interface the packet will be forwarded to. GPSRQ is based on per-hop behavior (Sect. 3.2.2) in a manner that the packet is directed toward nodes with satisfactory performance described by previously defined metrics and nodes which are geographically closer to the destination. According to the Differentiated Services Code Point (DSCP) fields transmitted in the IP packet header, GPSRQ also identifies the communication priority and determines the optimal path. The GPSRQ algorithm consists of two methods for forwarding packets: *greedy* and *recovery-mode* forwarding.

6.5.2 Greedy Forwarding

Assuming the existence of a service for the distribution of geographic information, it is then possible to form a table at each node with the latitudes and longitudes of other network nodes. Greedy routing involves forwarding packages to nodes which are geographically closest to the destination. Consider the example in Fig. 6.12, where an ingress node a, which is surrounded by three adjacent nodes k, b, and d, wants to send encrypted packets to egress node g. Ingress node a forwards the packet to b since the Euclidean distance between b and g is less than the distance between g and any of a's other neighbors. Greedy forwarding is repeated on interior nodes and ceases when the packet reaches its destination.

However, forwarding based only on Euclidean distance would lead to poor utilization of network links. Specifically, because the network nodes are assumed to be static nature, all traffic would always travel via the same links (in this case, link a-b-f), which would lead to the appearance of a bottleneck. Consequently, GPSRQ uses Eq. (6.16) to calculate the optimal path for forwarding packets:

$$F_{s,d,i} = (1 - \beta) \cdot R_{s,i} + \beta \cdot dist(i, d), \qquad (6.16)$$

where $dist$ is the Euclidean distance and R denotes the state of the link using Eq. (6.15), for each node i which belongs to the set N_s of all neighbors of source node s. Based on Eq. (6.16), all routes toward the destination are calculated and sorted in descending order. The route with the lowest value is preferred. The parameter β, which falls in the range [0,1], is used to optimize forwarding along the "geographically shortest" route or route with the most available resources. The parameter β allows a balance between different traffic classes providing the shortest paths for high priority traffic.

The routing method described above is effective if links to the destination are indeed available. However, one of the requirements in designing a routing protocol is high scalability and robustness, which means that alternative methods of defining a path must be found. Suppose the b-f link is congested or not enough keys are available to establish the path. Since no alternative path is available on node b, the packet will be returned to node a, which now needs to choose between its other neighboring nodes d and k. Node a, however, can obtain valuable information from the feedback received from node b. If the network packet could not be delivered to the destination via node b, i.e., if it was received over the same network interface from which it was previously sent, node a may conclude that a *routing loop* exists. The loop can be detected earlier and at node b. Node b will then raise the *loop* flag

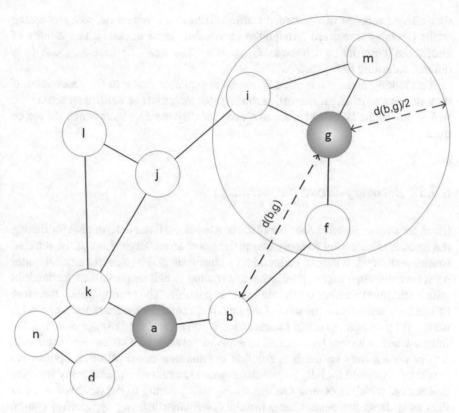

Fig. 6.12 Illustration of GPSRQ cache functionality. Ingress node *a* wants to send data to egress node *g*. After checking the availability of all routes on node *b*, the packet is returned to *a* if no available link is found. Node *a* then writes into the internal cache that it is not possible to route packets along path *a-b* toward the region indicated with a circle of radius $dist(b, g)/2$, with the center in node *g*. Any further request for routing toward any node which is placed in the defined circular region will be ignored over any route via node *b*, and an alternative route will be defined

in the GPSRQ packet header and return the packet to node *a*. When node *a* detects the *loop* flag, it will calculate the Euclidean distance $dist(b, g)$ and write into the internal cache memory that it is not possible to reach the packets located in the radius $dist(b, g)/2$) via node *b*. This robust caching mechanism enables acceleration of network traffic processing. If one packet has already reached the routing loop and lost time in delivery to the destination, there is no need for other packets in the network to travel the same route. Each subsequent query to route the packet to the node located in the defined circular region should also avoid node *b* and find an alternative route.

The question arises as to cache memory's validity, or more precisely, the definition of the optimal duration of records in memory. It primarily depends on the amount of traffic to be routed since it defines the number of queries and requests to establish a route and thus the number of records in the internal cache. It is assumed that a larger amount of forwarded traffic will have an impact on post-processing traffic (if both are realized through the same public infrastructure). The validity of entries can therefore be defined as $T_{cache} = \frac{1}{2} \cdot T_{maximal,b,g}$ where $T_{maximal,b,g}$ is defined according to Eq. (6.14).

This routing approach is valid if nodes are available closer to the destination. If they are not available, an alternative routing approach must be defined by setting the $inRec$ flag in the GPSRQ header to clearly identify the approach which should be used.

6.5.3   Recovery-Mode Forwarding

Since geographic information should not be used as the basic determinant for finding the route, it is necessary to apply alternative approaches. Regardless of the selected routing technique, it should be based on a consistent policy. Specifically, all nodes must have the same packet processing and routing to deliver packets to destinations with a minimum number of routing loops as possible. The routing policy therefore cannot be based only on the state of the link. If we consider the example in Fig. 6.12, node a could choose to direct packets toward node k and then toward node n if the links a-k and n-k were best ranked in terms of network link status. Although such a route choice may be the best solution in situations when all other options are exhausted, it should initially be avoided because it directs the packet away from the destination, which is contrary to the routing task (routing to the destination rather than away from the destination). Multiple alternative routing approaches can be applied. In an early paper, Karp suggested the use of the *right-hand rule*, which is based on forwarding so that the next edge from node a upon arriving from node b is the edge a-k, which is sequentially counterclockwise to edge a-b [27]. This type of routing remains in effect on all nodes (the $inRec$ flag is active in the GPSRQ header of the packet) until the packet arrives at the node where the greedy routing approach can be re-applied.

If the packet cannot reach a node, the danger exists that the packet will result in a routing loop. The loop can be eliminated by exchanging information about available other network nodes (an approach which has been deliberately avoided in GPSRQ routing because of the requirement to minimize the distribution of routing information as a result of the risk of passive eavesdropping). Another method is to pass the information concerning the nodes where the packet was processed in the package header. GPSRQ aims to reduce the amount of information in its headers, and uses only the IP address of the interface where the packet entered recovery-mode forwarding (Table 6.1).

Suppose the b-f and j-i links are unavailable due to a lack of a keys. Node a can send the packet to node b, which will return the packet. It will result in a cache entry on node a concerning the inability to route through node b. The packet is then routed to nodes k and j, which are geographically closest to destination g. Because the j-i link is unavailable, node j switches into recovery mode, marks the IP address of the j-l interface into the GPSRQ header, and directs the packet to node l (l is on the first edge counterclockwise about j from the line j-g, as required by the right-hand rule). The packet from l is routed again to k, forming a loop, and returns to node j. Node j reconsiders handling of the packet. It detects the IP address of its j-l interface in the GPSRQ header and concludes that the packet has been previously routed through that interface. Consequently, the packet cannot be redirected to the same j-l interface but must be returned to node k with the GPSRQ *loop* flag raised. Additionally, node j writes into its internal memory that it cannot reach g over the interface which leads toward node l. The cache entry contains three items: the IP address of the node l, the radius of circular region $dist(l,g)/2$, and value of the circle center set to the location of node g.

Node k adds a similar record to its cache memory. The record states that the destination cannot be reached via node j. The loop flag in the GPSRQ header is now set to 2, which indicates that the loop information has been processed successfully and an alternative path can be attempted. Node k then attempts node l, which once again forwards the packet toward j. Since j has no routing path other than the one on which the packet was received, it will set the *loop* flag to 1. After repeating the entire previously described procedure of adding the unavailable interfaces to the cache records, in the most extreme case, the packet will be returned to the source node a. If no route is available, the packet will be discarded at node a, but the records in the cache memories will remain to prevent the waste of network resources. Once the cache entry expires, the source node a can recommence the routing procedure.

Table 6.1 GPSRQ packet header fields used in recovery-mode forwarding

Field	Function
inRec	Forwarding mode: Greedy or recovery-mode
Loop	Returning loop indicator
recIF	IP address of the interface on which the packet entered recovery-mode

6.6 Summary

Because of the low rates in key generation, special attention must be given to how the keys are used. Taking into account that the distribution of keys consumes the same number of keys on each link through which distribution is performed (especially with OTP chipper), it is necessary to define and use the shortest paths. But, although the number of keys in key storage is the basic parameter dictating link availability in a QKD network, it is certainly not the only which affects link quality. Parameters such as delay and jitter during the delivery of keys are important to consider, especially when keys are used to protect real-time traffic. For specific reasons, applications may require dedicated requirements such as epsilon security or definitions of the vendor or type of device which generated the keys. All these parameters further affect the complexity of route calculation. This chapter described several methods of implementing QoS routing in a QKD network. They can be broadly categorized into two groups: one group uses metrics to describe links which contain multiple QoS parameters of interest and applies the well-known Dijkstra's algorithm or Bellman–Ford algorithm to the data collected in this manner; the other group adopts a heuristic view in which the shortest (by the number of hops) paths are additionally filtered according to QoS parameters. Routing in a QKD network may even be implemented with a combination of these concepts, but it primarily depends on how the network is organized and its QoS parameters are set.

References

1. Medhi, D., & Ramasamy, K. (2017). *Network routing: Algorithms, protocols, and architectures* (Vol. 51). ISBN 978-0-12-800737-2.
2. Cormen, T. H. (2013). *Algorithms unlocked*. MIT Press. ISBN 0-2625-1880-5.
3. Aguado, A., Lopez, V., Martinez-Mateo, J., Peev, M., Lopez, D., Martin, V., Lopez, V., Martinez-Mateo, J., Peev, M., Lopez, D., & Martin, V. (2017). GMPLS network control plane enabling quantum encryption in end-to-end services. In *2017 International Conference on Optical Network Design and Modeling (ONDM)*, Number 645127 (pp. 1–6), Budapest, Hungary. IEEE. ISBN 978-3-901882-93-7. https://doi.org/10.23919/ONDM.2017.7958519
4. Sasaki, M., Fujiwara, M., Ishizuka, H., Klaus, W., Wakui, K., Takeoka, M., Miki, S., Yamashita, T., Wang, Z., Tanaka, A., et al. Field test of quantum key distribution in the Tokyo QKD network. *Optics Express, 19*(11), 10387–10409.
5. Wang, S., Chen, W., Yin, Z.-Q., Li, H.-W., He, D.-Y., Li, Y.-H., Zhou, Z., Song, X.-T., Li, F.-Y., Wang, D., Chen, H., Han, Y.-G., Huang, J.-Z., Guo, J.-F., Hao, P.-L., Li, M., Zhang, C.-M., Liu, D., ... & Han, Z.-F. (2014). Field and long-term demonstration of a wide area quantum key distribution network. *Optics Express, 22*(18), 21739. ISSN 1094-4087. https://doi.org/10.1364/OE.22.021739
6. Xu, F., Chen, W., Wang, S., Yin, Z., Zhang, Y., Liu, Y., Zhou, Z., Zhao, Y., Li, H., Liu, D., Han, Z., & Guo, G. (2009). Field experiment on a robust hierarchical metropolitan quantum cryptography network. *Chinese Science Bulletin, 54*(17), 2991–2997.
7. Andersen, D. G., Snoeren, A. C., & Balakrishnan, H. (2003). Best-path vs. multi-path overlay routing. In *Proceedings of the 2003 ACM SIGCOMM conference on Internet measurement - IMC '03* (pp. 91). https://doi.org/10.1145/948205.948218

8. Zhu, Y., Dovrolis, C., & Ammar, M. (2006). Dynamic overlay routing based on available bandwidth estimation: A simulation study. *Computer Networks, 50*(6), 742–762. ISSN 13891286. https://doi.org/10.1016/j.comnet.2005.07.015

9. Gunkel, M., Wissel, F., & Poppe, A. (2020). Designing a quantum key distribution network - Methodology and challenges. *Photonische Netze - 20. ITG-Fachtagung* (pp. 49–51).

10. Rass, S., & König, S. (2012). Turning quantum cryptography against itself: How to avoid indirect eavesdropping in quantum networks by passive and active adversaries. *International Journal on Advances in Systems and Measurements, 5*(1), 22–33.

11. Kollmitzer, C., & Pivk, M. (2010). *Applied quantum cryptography* (Vol. 797). Springer Science & Business Media. ISBN 3642048293. https://doi.org/10.1007/978-3-642-04831-9

12. Konig, S., & Rass, S. (2011). On the transmission capacity of quantum networks. *International Journal of Advanced Computer Science and Applications(IJACSA), 2*(11), 9–16. https://doi.org/10.1.1.671.7300

13. Mehic, M., Niemiec, M., Rass, S., Ma, J., Peev, M., Aguado, A., Martin, V., Schauer, S., Poppe, A., Pacher, C., Voznak, M., Ma, J., Peev, M., Aguado, A., Martin, V., Schauer, S., Poppe, A., Pacher, C., & Voznak, M. (2020). Quantum key distribution : A networking perspective. *ACM Computing Surveys, 53*(5). ISSN 15577341. https://doi.org/10.1145/3402192

14. Martin, V., Brito, J. P., Escribano, C., Menchetti, M., White, C., Lord, A., Wissel, F., Gunkel, M., Gavignet, P., Genay, N., Le Moult, O., Abellán, C., Manzalini, A., Pastor-Perales, A., López, V., & López, D. (2021). Quantum technologies in the telecommunications industry. *EPJ Quantum Technology, 8*(1), 19. ISSN 2662-4400. https://doi.org/10.1140/epjqt/s40507-021-00108-9

15. Braden, R. T., Clark, D. D., & Shenker, S. (1994). *Integrated Services in the Internet Architecture: an Overview*. RFC 1633. https://rfc-editor.org/rfc/rfc1633.txt

16. Dianati, M., Alleaume, R., Gagnaire, M., Shen, X. (Sherman). (2008). Architecture and protocols of the future European quantum key distribution network. *Security and Communication Networks, 1*(1), 57–74. ISSN 19390114. https://doi.org/10.1002/sec.13

17. Huitema, C. (1995). *Routing in the internet*. Prentice-Hall.

18. Li, M., Quan, D., & Zhu, C. (2016). Stochastic routing in quantum cryptography communication network based on cognitive resources. In *2016 8th International Conference on Wireless Communications and Signal Processing, WCSP 2016*. https://doi.org/10.1109/WCSP.2016.7752564

19. Mehic, M., Fazio, P., Voznak, M., & Chromy, E. (2016). Toward designing a quantum key distribution network. *Advances in Electrical and Electronic Engineering, 14*(4Special Issue), 413–420. ISSN 18043119. https://doi.org/10.15598/aeee.v14i4.1914

20. ITU-T. (2002). *Recommendation M.2301: Performance objectives and procedures for provisioning and maintenance of IP-based networks*. Technical Report, ITU-T, 2002. https://www.itu.int/rec/T-REC-M.2301-200207-I/en

21. Elliott, C., Colvin, A., Pearson, D., Pikalo, O., Schlafer, J., & Yeh, H. (2005). Current status of the DARPA quantum network (Invited Paper). In E. J. Donkor, A. R. Pirich, & H. E. Brandt (Eds.), *Proceedings of the SPIE 5815, Quantum Information and Computation III,* (Vol. 5815, pp. 138–149). https://doi.org/10.1117/12.606489

22. Chip Elliott and H Yeh. *DARPA Quantum Network Testbed*. Technical Report July, BBN Technologies Cambridge, New York.

23. Pearson Brig, B., & Elliott Spencer, D. (2010). *Systems and methods for implementing routing protocols and algorithms for quantum cryptographic key transport*. http://www.google.com/patents/US7706535

24. Yang, C., Zhang, H., & Su, J. (2017). The QKD network: model and routing scheme. *Journal of Modern Optics, 64*(21), 2350–2362. ISSN 13623044. https://doi.org/10.1080/09500340.2017.1360956

25. Ma, C., Guo, Y., Su, J., & Yang, C. (2016). Hierarchical routing scheme on wide-area quantum key distribution network. In *2016 2nd IEEE International Conference on Computer and Communications (ICCC)* (Vol. 1, pp. 2009–2014). IEEE. ISBN 978-1-4673-9026-2. https://doi.org/10.1109/CompComm.2016.7925053

26. Mehic, M., Fazio, P., Rass, S., Maurhart, O., Peev, M., Poppe, A., Rozhon, J., Niemiec, M., & Voznak, M. (2020). A novel approach to quality-of-service provisioning in trusted relay quantum key distribution networks. *IEEE/ACM Transactions on Networking, 28*(1), 168–181. ISSN 1063-6692. https://doi.org/10.1109/TNET.2019.2956079

27. Karp, B., & Kung, H. T. (2000). GPSR: Greedy perimeter stateless routing for wireless networks. In *Proceedings of the 6th Annual International Conference on Mobile Computing and Networking - MobiCom '00* (pp. 243–254), New York, New York, USA: ACM Press. ISBN 158-11-3197-6. https://doi.org/10.1145/345910.345953

28. Mehic, M., Maurhart, O., Rass, S., Komosny, D., Rezac, F., & Voznak, M. (2017). Analysis of the public channel of quantum key distribution link. *IEEE Journal of Quantum Electronics, 53*(5), 1–8.

29. Paris, S., Nita-Rotaru, C., Martignon, F., & Capone, A. (2013). Cross-layer metrics for reliable routing in wireless mesh networks. *IEEE/ACM Transactions on Networking, 21*(3), 1003–1016. ISSN 10636692. https://doi.org/10.1109/TNET.2012.2230337

30. Li, J., Jannotti, J., De Couto, D. S. J., Karger, D. R., & Morris, R. (2000). A scalable location service for geographic ad hoc routing. In *Proceedings of the 6th Annual International Conference on Mobile Computing and Networking - MobiCom '00* (pp. 120–130). https://doi.org/10.1145/345910.345931

31. Schmitt-Manderbach, T., Weier, H., Fürst, M., Ursin, R., Tiefenbacher, F., Scheidl, T., Perdigues, J., Sodnik, Z., Kurtsiefer, C., Rarity, J. G., Zeilinger, A., & Weinfurter, H. (2007). Experimental demonstration of free-space decoy-state quantum key distribution over 144 km. *Physical Review Letters, 98*(1), 010504.

32. Sheikh, K. H., Hyder, S. S., & Khan, M. M. (2006). An overview of quantum cryptography for wireless networking infrastructure. In *Proceedings of the 2006 International Symposium on Collaborative Technologies and Systems, CTS 2006* (pp. 379–385). https://doi.org/10.1109/CTS.2006.16

33. Vallone, G., Bacco, D., Dequal, D., Gaiarin, S., Luceri, V., Bianco, G., & Villoresi, P. (2015). Experimental satellite quantum communications. *Physical Review Letters, 115*(4), 040502. ISSN 0031-9007. https://doi.org/10.1103/PhysRevLett.115.040502

34. Wijesekera, S. (2011). *Quantum Cryptography for Secure Communication in IEEE 802 . 11 Wireless Networks*. PhD Thesis.

35. Basagni, S., Chlamtac, I., Syrotiuk, V. R., & Woodward, B. A. (1998). A distance routing effect algorithm for mobility (DREAM). In *Proceedings of the 4th Annual ACM/IEEE International Conference on Mobile Computing and Networking - MobiCom '98* (pp. 76–84). New York, New York, USA: ACM Press. ISBN 1-581130-35-X. https://doi.org/10.1145/288235.288254

Chapter 7
From Point-to-Point to End-to-End Security in Quantum Key Distribution Networks

Because quantum key distribution is a technology for establishing keys for symmetric encryption (preferably one-time pads, but more practically keys for AES or other conventional symmetric cryptography), end-to-end confidentiality or authentication requires a deeper look into network structure and comes with additional assumptions.

7.1 Single-Path Transmission: Trusted Relay

The simplest method of accomplishing end-to-end security from point-to-point security in QKD is "by assumption". Let us consider a message m which enters the network at Alice's computer. Alice shares a QKD key k_1 with its first (logically) connected node v_1 along a path to receiver Bob; let the entire path be Alice $\rightarrow v_1 \rightarrow v_2 \rightarrow \cdots \rightarrow v_n \rightarrow$ Bob. Relay node v_1 in this network is not necessarily a device with which Alice shares a cable, but it must be a device where Alice can run a QKD protocol or key $k_{A,1}$ can be securely exchanged by other means. This must hold for all consecutive relays along the path, i.e., the connection $v_i \rightarrow v_{i+1}$ is secured with a key $k_{i,i+1}$ for $i = 1, 2, \ldots, n$, where the end-nodes Alice and Bob own the keys $k_{A,1}$ and $k_{n,B}$.

For now, let us assume that an infinite amount of quantum key is available between any two relays which run QKD protocols endlessly to (re)generate key material. Alice can then conceal her message beneath a one-time pad $k_{A,1}$ to send $c_1 = m \oplus k_{A,1}$ to the first relay v_1. This relay can decrypt and re-encrypt the payload for the next hop v_2 by first uncovering the plain text via an XOR $m \leftarrow c_1 \oplus k_{A,1}$ and forwarding $c_2 \leftarrow m \oplus k_{1,2}$ to the next relay v_2. This process continues likewise inside v_2, where c_2 is received and forwarded $c_3 = c_2 \oplus k_{2,3}$, and so on, until Bob receives the final ciphertext $c_B = k_{n,B} \oplus m$ for decryption using the final key $k_{n,B}$ shared with his own access point v_n to the quantum network. Figure 7.1

© Springer Nature Switzerland AG 2022
M. Mehic et al., *Quantum Key Distribution Networks*,
https://doi.org/10.1007/978-3-031-06608-5_7

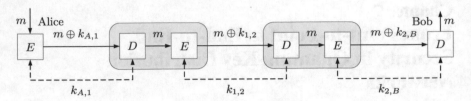

Fig. 7.1 Single path transmission based on a one-time pad

illustrates the special case of transmission of m over two hops from Alice to Bob. The functions D and E in this diagram are general decryption and encryption functions, implemented as $D(c, k) = c \oplus k$ and $E(m, k) = m \oplus k$ for a one-time pad, but may be replaced by any other encryption, such as AES as a temporary fallback encryption, if the nodes run out of key material [1] before the buffers can be refilled.

The conceptual simplicity comes with the price tag of uncovering the message m within each relay node v_i on the path to Bob so that a successful hack of any among $v_1, \ldots, v_n$ will eventually make the entire transmission insecure, despite the unconditional strength of the one-time pad and quantum technology. The assumption that such a hack *does not occur* is known as a *trusted relay*: in essence, this entails two things:

1. No eavesdropping occurs on the link, as assured by the QKD protocols.
2. *No eavesdropping occurs within the forwarding relay*, since this device is under strong technical protection.

Protection of the devices from unauthorized physical or logical access is a matter beyond quantum technologies and entails an entire spectrum of organizational access control and restriction mechanisms which are well known from many other high-security domains. Purely quantum-based protection would—theoretically—be possible using quantum repeaters, but even that case may not work reliably without any physical access protection.

Practically, matters are somewhat more involved, because the devices will still have to run firmware whose very complexity opens it up to its own attack surface, including all the software bug possibilities, such as buffer overflows, return-oriented programming, etc. Even if established under physical protections, access over maintenance interfaces or remote configuration may be necessary, which potentially opens channels into the memory areas where the secret payloads pass through. We shall not dive into the complex matters of security management here, other than pointing out that QKD devices, in a manner, lend themselves to similar risk and security management measures to any other enterprise asset with high security clearance. Therefore, setting a QKD device to work under the trusted relay assumption needs classic asset and risk management, for example, using ISO27k, ISO31k, BSI Standards or others.

Besides the requirement of security risk management to exist independently and as an umbrella to a QKD network in any case, it is easy to argue that the trusted node assumption is not avoidable, unless we are willing to drop at least one of the distinguishing features of QKD:

- Alice and Bob share no common presupposed secret: if they did, then Alice could directly use her secret to symmetrically encrypt message m for Bob, and the transmission would simply be delivery of an encrypted payload. Strictly speaking, this assumption is not even entirely correct, since to run QKD, both end-points need a (short) common secret to exist from the beginning simply to authenticate their protocol and to avoid person-in-the-middle attacks.

  To avoid falling back to "computational security only", Alice and Bob would need the ability to generate arbitrary amounts of end-to-end shared secrets, which either requires a direct QKD connection (not assumed), or other means of efficient key exchange. Because Alice shares only a fixed amount of available key material directly with Bob, all they can do is use this (short) key to encrypt their payload, essentially reverting to what classical cryptography can do equally well (e.g., hybrid encryption).

- Alice and Bob want their security to be unconditional: removing this requirement would also defeat the entire purpose of QKD, because if computational security is enough, then post-quantum cryptography is an alternative to achieving end-to-end security by means of public-key or symmetric encryption. This is indeed an option to bridge short-term DoS attacks on links by eavesdropping: if an eavesdropper is detected, the current key material is abandoned for being insecure, and Alice and Bob can continue their conversation by using stored key material from earlier periods (when the link was not eavesdropped). But since the eavesdropper is currently present, they will be unable to refill their local key-stores, which will eventually run dry. Before this occurs, they may switch to classic encryption and use the same and constant number of keys, for example, to AES-encipher the payload instead of using one-time pads. This temporary trade of perfect confidentiality for continuous availability lets us avoid a DoS but not relieve the trusted node assumption if the overall transmission regime remains unchanged.

Remark 7.1 Without digressing into an ethical debate here, we note that trusted relay can indeed be a desired feature for law enforcement authorities against criminals within the network while remaining thoroughly protected against external access. For example, if the quantum network is a service provided and maintained by a government authority, trusted relay will keep all transmissions within the network open to this government authority yet keep transmission completely hidden from the public and equally well protected against foreign espionage from external attackers. Although this may be an independent appeal of trusted relay over stronger implementations using multiple paths or randomized routing, we retain this accessibility, yet make its exploitation more difficult.

Adopting an economic perspective on the problem of hacking a QKD network, a rationally acting intruder would probably not bother much with breaking into a link that is unconditionally protected by the laws of quantum physics. Instead, the node, running some (possibly well known) operating system is the *much easier target*, and attacks may be most likely expected on the nodes, not on the QKD links. Securing the nodes is a matter of traditional security management, such as a technical measure which applies several instead of only one path to send a message.

7.2   Relaxing the Trust Assumption: Multipath Transmission

Since we cannot hope for a zero-trust solution regarding the QKD relay nodes, the alternative is to reduce the amount of information flowing through a node so that even if a node is hacked, perhaps not all information is lost, only a part of it.

To formalize this idea and operationalize the concept, let us view the network as a graph $G = (V, E)$, with V containing all nodes in the network, such as routers, switches, etc., and $E \subseteq V \times V$ being the set of edges in the graph, corresponding to physical links (cables, but also wireless connections) between nodes in G. Unless stated otherwise, we assume all connections to be bi-directional, thus G is an undirected graph. An *s—d-path*, or simply just *path*, in G from a *sender* or *source* $s \in V$ to a destination $d \in V$ is an ordered sequence of edges, denoted π and expressed as $\pi = (e_1, e_2, \ldots, e_n)$, with all $e_i \in E$, and $e_1 = (s, v_1), e_n = (v_n, d)$. The number n is called the *length* of a path, and is denoted $|\pi|$. In the network realm, this would be the number of "hops" that a message takes from s to d. The symbol $V(\pi)$ is the set of all nodes along the path (and we have the length $|\pi| = |V(\pi)|$). In Fig. 7.2, we have $s = $ Alice and $d = $ Bob. A path, among several, from Alice to Bob appears as $\pi_1 = ((s, 4), (4, 5), (5, d))$ or $\pi_2 = ((s, 1), (1, 2), (2, d))$, etc. From here onward, it would be convenient to denote paths with a simpler notation, for example, $\pi_1 = s$—4—5—d or $\pi_2 = s$—1—2—d.

Let us represent a quantum network as an undirected graph G, with the additional assumption on the edge set declaring that once a pair $u, v \in V$ of nodes shares an edge $e = \{u, v\}$, the two run a QKD protocol between them. The trusted relay assumption manifests in this formalization as the hypothesis that eavesdropping

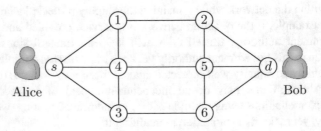

Fig. 7.2 Example network topology (adapted from [2])

- *cannot occur* anywhere on E, because the edges are perfectly protected by QKD, and
- *will not occur* anywhere on V, because the nodes are "trusted", i.e., robust and able to resist hacking.

Our goal now is to dispense with the second assumption, or at least relax it as far as possible. It turns out that such relaxation is possible, but at the cost of trading it for a stronger assumption on the network topology.

7.2.1 Quantifying the Probability of Eavesdropping

Let us suppose that the quantum network has a topology that allows for $k \geq 2$ paths between any two nodes. The simplest instance is a ring topology, providing exactly two routes between any two points in the network. This is a common practice in the layout of many practical networks for the simple reason of safety against failure of a single node or cable (e.g., in the case of maintenance or replacement).

We can use such redundancy not only to increase reliability but also boost confidentiality if we route a packet over the two paths. Again, in the simplest instance, the sender could take a message m and "decompose" it into two random strings s_1, s_2 by virtue of picking a random string $r \in \{0, 1\}^{|m|}$ and inputting $s_1 := m \oplus r, s_2 := r$. If the two routes π_1, π_2 in the network graph are available, then transmitting s_1 over π_1 and s_2 over π_2, assuming a trusted relay, enforces the adversary to intercept both π_1 and π_2, accordingly, to recover the message.

Obviously, this method doubles network traffic, which is the price of increasing the difficulty for the adversary having to intercept two lines by hacking at least two nodes (and not any two nodes, but two nodes on distinct paths in G).

The concept can be generalized in various ways since it is conceptually straightforward to divide the message into more than two pieces by picking random strings $r_1, r_2, \ldots, r_{t-1} \in \{0, 1\}^{|m|}$ and transmitting $s_1 = r_1, s_2 = r_2, \ldots, s_{t-1} = r_{t-1}$ and $s_t = m \oplus r_1 \oplus r_2 \oplus \cdots \oplus r_t$ over distinct paths in the network, assuming that the topology is sufficiently meshed to allow this. Suppose we have found several paths $\pi_1, \pi_2, \ldots, \pi_\ell$, where the number ℓ is generally exponential in the node count $|V|$ but will in practice be around $\ell = 2$ for a circular or only slightly larger network layout.

In the best possible case, we have $\ell \geq t$ paths from the sender s to the destination d, all of which meet only at s and d; formally expressible as $V(\pi_i) \cap V(\pi_j) = \{s, d\}$ for all $i \neq j$. We call two such paths *node-disjoint*. A well known theorem by H. Whitney relates the existence of node-disjoint paths to the vertex connectivity of graphs: a graph $G = (V, E)$ is said to have *vertex connectivity* ℓ if it remains connected upon deletion of up to ℓ nodes. This property directly states how many node-disjoint paths exist:

Theorem 7.1 (Whitney [3, Thm. 5.17]) *A nontrivial graph G is ℓ-connected for some integer $\ell \geq 2$, if and only if for each pair u, v of distinct vertices of G at least ℓ node-disjoint u—v paths exist in G.*

Ring topologies have vertex connectivity 2; removing any node opens the ring but leaves a line with all nodes on it intact so that all nodes can still communicate. Star topologies and trees have vertex connectivity 1; removing the central node from a star or any node from a tree will leave at least two nodes unable to communicate. The network in Fig. 7.2 has vertex connectivity 2; any removal of up to two nodes leaves Alice and Bob connected but removing three nodes (e.g., $\{1, 4, 3\}$) disconnects them.

While the computation of the vertex connectivity number is generally NP-hard [4], it is surprisingly simple to inductively construct networks with a desired connectivity. Here, we will not further digress, but we return to this in Sect. 7.2.2.

Let us be optimistic from the beginning and assume the existence of $\ell \geq t$ node-disjoint paths between the sender s and the destination node d in the graph. Then, we could transmit each part $s_1, \ldots, s_t$ concurrently over its own path from s to d, with the assurance that the attacker will not learn anything about the message unless it manages to intercept all t paths simultaneously. We have therefore practically relaxed the trusted node assumption into the weaker hypothesis that confidentiality is guaranteed unless more than t nodes fall victim to a hacker attack simultaneously.

Now, toward a more realistic expectation, let us suppose the existence of less paths than message parts, therefore to transmit $t > \ell$ parts, we must send two or more parts over the same path sequentially. The situation is now similar but more complex to analyze, because whether the adversary can disclose the message depends on whether it is capable of catching all t message parts from within the relay nodes. For example, if two concrete paths π_1, π_2 intersect at some common node $v \in V(\pi_1) \cap V(\pi_2)$, e.g., an entry or exit gateway to some subnetwork, hijacking this node v is apparently more valuable to the attacker than targeting a node through which only one path proceeds.

This brings us to the question of which nodes may be under particular risk of hijacking, as opposed to other nodes which are under stronger protection and hence less likely to become attacked. Answering this question is a matter of pondering the adversary's assumed capabilities and the vulnerabilities that exist in the network. Identifying weak spots in the network is called Topological Vulnerability Analysis (TVA) [5] and typically outputs detailed reports that annotate all network links and devices with vulnerabilities obtained by searching public databases (e.g., the National Vulnerability Database (NVD) [6] maintained by the National Institute of Standards and Technology (NIST)) concerning the respective devices. This information is often augmented with numeric scores according to schemes such as Common Vulnerability Scoring System (CVSS) [7], or others.

Matching the assumed capabilities with the details of the vulnerabilities found in the network then raises a set of possible subsets of nodes which the assumed attacker can realistically conquer. For example, if all devices run on the same firmware with a known exploitable software bug, these devices would be listed as points where an intruder can gain entry to the network and drain secret information during re-encryption in the trusted relay. The collection of all such sets forms a so-called *adversary structure*, formally being a subset $\mathscr{A} \subseteq \mathscr{P}(V)$ of the powerset of the graph's node set. Theoretically, $\mathscr{A}$ can be any collection of subsets and in practice will be composed of nodes which all have one or more vulnerabilities (with known exploits) in common. Without an explicit reference to vulnerabilities, we could also alternatively define the adversary structure on the assumption that "no more than t nodes can be hacked", establishing the setting that $\mathscr{A} = \{A \subset V : |A| \leq t\}$. This has the appeal of sparing the need to identify and model vulnerabilities, but it is computationally more challenging to manage, as the size of $\mathscr{A}$ is in the order of $O(|V|^t)$ and may quickly become intractable when t becomes large.

Returning to our original intention of quantifying the probability of eavesdropping, let us revise the ingredients available so far:

- A list $L = \{\pi_1, \pi_2, \ldots, \pi_\ell\}$ of paths between s and d, node-disjoint where possible, but permitted to have nodes in common.
- A collection $\mathscr{A} = \{A_1, A_2, \ldots\}$ of sets we assume is possibly under the attacker's control due to common vulnerabilities or other conditions.

The maximum sizes of ℓ and $|\mathscr{A}|$ are theoretically of exponential magnitude in a number $|V|$ of nodes, but practically, we may obtain L as a relatively short list (typical network topologies will not admit more than $\ell = 2$ paths) and $\mathscr{A}$ as a reasonably sized collection of vulnerabilities identified during a TVA, assuming that these are all that an attacker can exploit (at this point, let us postpone handling zero-day risks until later). We refer the reader to [8, 9] for a study and suggestions on how to do this practically.

With these two inputs, we can set up a tableau (a matrix), judging each possible attack indicated by $\mathscr{A}$ against each possible transmission according to L.

Sending the Message Over a Single Random Path

In the simplest case, and avoiding multiplication of the network load by the factor $t > 1$, we may select a route $\pi \in L$ at random and send the message over this route, assuming a trusted relay. The adversary, in turn, now needs to correctly "guess" the node(s) to intercept and read m from within (Fig. 7.1).

This results in a tableau as follows:

	A_1	$\cdots$	A_j	$\cdots$
π_1	u_{11}	$\cdots$	u_{1j}	$\cdots$
$\vdots$		$\ddots$		
π_i	u_{i1}	$\cdots$	u_{ij}	$\cdots$
$\vdots$	$\vdots$		$\vdots$	$\ddots$

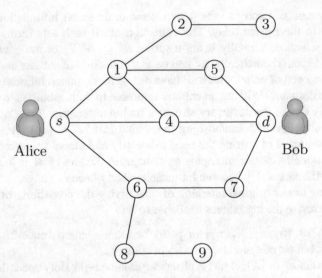

Fig. 7.3 Network as a transmission game playground (adapted from [2])

in which the entries u_{ij} are indicators which state whether the message can be disclosed. For an assumed trusted relay, we put

$$u_{ij} = \begin{cases} 0, \text{ if the message was disclosed, i.e., if } V(\pi_i) \cap A_j \neq \emptyset; \\ 1, \text{ otherwise, since the message bypassed all hijacked nodes.} \end{cases} \quad (7.1)$$

This results in a 0–1-valued matrix.

Example 7.1 (Adapted from [2]) If Alice uses only a single transmission line, the network as shown in Fig. 7.3 admits four paths, namely:

1. $\pi_1 = s$—4—t
2. $\pi_2 = s$—1—4—d,
3. $\pi_3 = s$—1—5—d, and
4. $\pi_4 = s$—6—7—d.

The attacker, however, can strike everywhere, totalling 9 points, excluding Alice's and Bob's nodes, for obvious reasons.

The resulting matrix is 4×9, and found as

↓path	Attacked node								
	1	2	3	4	5	6	7	8	9
π_1	1	1	1	0	1	1	1	1	1
π_2	0	1	1	0	1	1	1	1	1
π_3	0	1	1	1	0	1	1	1	1
π_4	1	1	1	1	1	0	0	1	1

As evident from the network and how the paths are defined, an attacker has no incentive to strike nodes 2, 3, 8 or 9 since these will carry no message flows. In the matrix, their corresponding columns appear all with 1, indicating that Alice and Bob will surely exchange their content in privacy. But if Alice uses π_1, then hacking either node 1 or node 4 will disclose the data in the case of a trusted relay. Similarly, if path π_3 is selected by either Alice or the routing system, the attacker will obtain the message if it can intercept (hack) nodes 1 or 5.

Toward a quantification of secrecy, let us consider the resulting tableau as the payoff structure for a mathematical game between two players:

Player 1 is the sender, who seeks to transmit the message in confidentiality to the destination node d.

Player 2 is the adversary, who seeks to disclose the secret message.

In this view, we can ask for the Nash equilibrium of the resulting zero sum game, which entails computing three results:

1. An optimal choice distribution over the paths in L to minimize any chance of the message becoming disclosed,
2. A likewise optimal choice over the attacker's options in $\mathscr{A}$ to maximize its chances of discovering the secret message,
3. A saddle point value, giving the likelihood of message disclosure.

The third output is the sought quantification, and formally, arrives at the solution of the following min-max optimization problem:

$$v = \max_{p \in \Delta(L)} \min_{q \in \Delta(\mathscr{A})} x^T \cdot U \cdot y, \tag{7.2}$$

where Δ is the set of (categorical) probability distributions over the sets L, and respectively, $\mathscr{A}$, and $U = (u_{ij})$ is the matrix, i.e., tableau from above. An element $p \in \Delta(L)$ is a function that takes a path $\pi \in L$ and outputs the probability $p(\pi)$ of selecting this path in a randomized routing. The problem for player 1 in the above game is to find the *optimal* probabilities for this randomized choice. Likewise, each element in $\Delta(\mathscr{A})$ is a function that takes a certain set $A \in \mathscr{A}$ of nodes (e.g., characterized by a certain exploit mountable on physical devices) and returns the probability $q(A)$ to attempt this exploit when mounting an attack. The problem for adversarial player 2 is to determine the most likely, i.e., the most promising, attacks to minimize the expectation of a secret transmission, which is

$$x^T \cdot U \cdot y = E_{i \sim p, j \sim q}(u_{ij})$$

$$= \sum_{\pi \in L, A \in \mathscr{A}} p(L) \cdot q(A) \cdot u_{ij}$$

$$= \Pr(\text{message remains confidential}),$$

where the equality of the expectation with the probability is because we have defined the tableau above with indicator variables (only) and by observation that the probability of an event is (defined as) the expected value of some indicator variable.

That said, Eq. (7.2) can be rephrased as the statement that v equals the likelihood for a secrecy breach, given that:

- The sender does its best (maximize over all path choices) to transmit the message in privacy.
- The attacker does its best (minimize over its own choices) to disclose the message.

This is essentially the concept of a Nash equilibrium, and the solution to (7.2) delivers the value v attained at the *simultaneous optima* p^* for the sender and q^* for the attacker. These two are the remaining results from above, namely, the optimal choice distribution for randomized routing and the optimal choice distribution for where to attack.

Let us postpone the algorithmic details of computing v, p^* and q^* until Sect. 7.2.1.1 and first examine this analysis more closely: the quantitative security obtained from this analysis is indeed the value v, with the interpretation that

$$v = \Pr \left(\begin{array}{l|l} \text{message is transmitted} & \text{random routing according to } p^*, \\ \text{in perfect secrecy} & \text{and random attacks according to } q^* \end{array} \right).$$

Example 7.2 (Example 7.1 Continued) Using the previous matrix, we first note that we can safely abandon all columns containing 1, since these correspond to attack strategies which have no payoff for the attacker. In game-theory jargon, we would call these columns *dominated*. Likewise, we only need to keep one copy of each column. For example, it makes no difference to the attacker whether it strikes node 6 or 7, regardless of the path Alice chooses. Such columns are called *weakly dominated*, and their deletion does not invalidate or alter the value of v which we compute.

What remains is a smaller 4×4-matrix (thus reducing the complexity for analysis) of the form:

	Attacked node			
↓path	1	4	5	6
π_1	1	0	1	1
π_2	0	0	1	1
π_3	0	1	0	1
π_4	1	1	1	0

The value v, determined with the methods described in Sect. 7.2.1.1, is

$$v = \frac{2}{3},$$

attained at the optimal randomized routing rule to select

- path π_1, π_3 or π_4, equiprobable with the probability $1/3$,
- and never select path π_2.

Similarly, the attacker obtains an optimal attack pattern to strike nodes 1, 4 or 6 with uniform probability $1/3$ and spare attacking node 5 (attack probability zero).

These numbers, as far as they concern advice for the attacker, need to be handled with care by Alice and Bob, because if these two discover that node 5 has no likelihood of ever being hacked, this *does not* mean no existence of another equilibrium which assigns a positive likelihood to node 5. The optima existing in (7.2) are generally non-unique, and either only one, a finite number or even an infinitude can exist. Any optimum is equally good for the player it refers to, but non-unique for the respective opponent. Therefore, even if Alice obtains a best strategy for her transmission, she *cannot use* the optimum she computes for the attacker to inform her which nodes will most likely be assailed. Symmetrically, the attacker has *one* possibility for Alice to transmit optimally, yet no guarantee that Alice will conform to this.

The conditionality of v on the sender implementing a randomized routing regime according to some choice distribution p^* over the paths is technically possible (even if not trivial), but we can hardly rely on the attacker to nicely follow the assumption of attacking at random points and as prescribed by q^*. However, since q^* is actually the *best* that the attacker can do, this means that anything other than attacking at random according to q^* would only *increase* the sender's chances to transmit the message in privacy. More rigorously, and drawing somewhat more from game theory (not discussed any further), we can prove that

$$\Pr\left(\text{message is disclosed} \,\middle|\, \begin{array}{c}\text{random routing according to } p^*, \\ \text{and } \textit{arbitrary} \text{ attack patterns} \\ \text{based on } \mathscr{A}\end{array}\right) \leq 1 - v.$$

$$(7.3)$$

This is already much stronger than the previous interpretation, because it tells us that no matter what the adversary does, as long as no unexpected attack occurs, i.e., something *not covered* by $\mathscr{A}$, the value $1 - v$ is a guaranteed upper bound of the probability of the message becoming disclosed.

The main output artifacts of the game theoretic analysis, namely the Nash equilibrium for the defender, and the saddle point value of the game, then have a natural interpretation as SLA and Operational Level Agreement (OLA):

- The value $1 - v$ by virtue of (7.3) is a guarantee for the defender regarding a certain degree of confidentiality, quantified as a likelihood. This is nothing else than an SLA, once we present it to a customer.

- Similarly, the optimal strategy for the defender to accomplish the guarantee of (7.3) is a prescription of how to use the network to this end. This is nothing else than an OLA.

Investigating the conditioning in (7.3), we may ask what happens if the attacker somehow acts unexpectedly and strikes at a point we did not expect according to $\mathscr{A}$. Technically, this is a case when our adversary model $\mathscr{A}$ is incorrect, and practically, we call this an occurrence of a *zero-day attack*.

In the above setting, and in the jargon of game theory, this corresponds to a "winning strategy" in a game which zeroes the expectations of player 1. With the indicators u_{ij} as defined in (7.1), the winning strategy would appear in the tableau as a column of all zeroes, and the resulting optimum for the adversary is selecting exactly this column regardless of the sender's actions (it will always result as a disclosure of the message). This column, as an attack strategy, is said to *dominate* all other strategies. Depending on the network topology, winning strategies for the attacker may also arise from points of weak connectivity, i.e., single nodes through which all traffic needs to flow. Such nodes can be entry or exit gateways, access points into the last mile toward the customer, or other hubs in the network. Any such "sweet spot" may require special protection, since otherwise, it may present itself as a winning strategy for the attacker in the game. Let us illustrate this by modifying the previous example accordingly:

Example 7.3 Consider the network topology displayed in Fig. 7.4.

Suppose that Bob connects via node 1, which is his access point to the network. Obviously, assuming a trusted relay, the most promising target for an attacker would be breaking into node 1 to read all of Bob's traffic. And indeed, modeling this transmission as a game matrix reveals the weakness as a column of all zeroes:

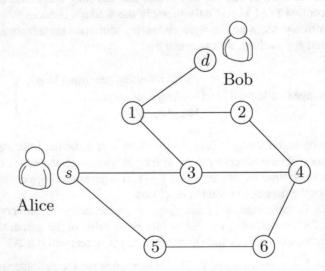

Fig. 7.4 Network with a vulnerable spot between Alice and Bob

↓path	Attacked node					
	1	2	3	4	5	6
s—3—1—d	0	1	0	1	1	1
s—3—4—2—1—d	0	0	0	0	1	1
s—5—6—4—3—1—d	0	1	0	0	0	0
s—5—6—4—2—1—d	0	0	1	0	0	0

Recomputing the value v as previously will yield $v = 0$, indicating zero-chance for Alice to transmit her message to Bob in privacy. This is consistent with the network layout and the assumption that Bob's access point can be hacked.

Now, suppose that we *harden* node 1 accordingly. If an attack is made or assumed impossible on this node, it amounts to removing the column from the analysis. Looking at the now reduced tableau, we have:

↓path	Attacked node				
	2	3	4	5	6
s—3—1—d	1	0	1	1	1
s—3—4—2—1—d	0	0	0	1	1
s—5—6—4—3—1—d	1	0	0	0	0
s—5—6—4—2—1—d	0	1	0	0	0

This time, we compute the value $v = 1/2$ and best attempts for Alice to send her data over the path s—3—1—d or path s—5—6—4—2—1—d, both with a probability of 1/2. The attacker's best action is to strike nodes 2 or 3 with uniform probability (we remark that the same result would be achieved if we exclude one of the last two columns, as both attack strategies satisfy the attacker identically).

The method as outlined does not lend itself to a quantification of the likelihood of such a situation (a zero-day exploit) occurring, but *we can resolve* this situation once we become aware of it. Indeed, we can show that given a dominating column, it is sufficient to *resolve only one scenario* to remove the column's dominance over others. Such a change takes us from $v = 0$ to $v > 0$, such that $1 - v < 1$ in (7.3) [10]. Once this is accomplished, we can make the bound in (7.3) arbitrarily small by repeating the protocol as often as we can afford to bring down the probability of the message becoming disclosed arbitrarily close to zero. Let us continue the previous example by assuming that node 1 remains vulnerable, but that a small chance exists of the attacker missing the traffic when it passes through node 2 (for whatever reason).

The game model of this transmission is the same as above, but it has an entry $p > 0$ at some position in the first column so that this column is no longer dominated (in the strategic sense of game theory). Let us suppose that the tableau is now:

↓path	Attacked node					
	1	2	3	4	5	6
s—3—1—d	$p = 0.2$	1	0	1	1	1
s—3—4—2—1—d	0		0	0	0	1	1
s—5—6—4—3—1—d	1		1	0	0	0	0
s—5—6—4—2—1—d	0		0	1	0	0	0

Recomputing the value v, we now obtain $v = 1/6$ and can restore security. Likewise, if we input some value $p > 0$ into any other row of the first column, we would find the respective values $v > 0$ in all four cases, illustrating that *any* countermeasure to the winning strategy would be enough to regain at least some security.

The detail which should be highlighted here is:

A security fix may not need to address all possibilities! It can indeed be enough to fix some (at least one) scenarios to restore the possibility of perfectly secure end-to-end communication.

Having restored v to the value > 0, this may still be small. The next question is how to raise the security level, in the best case, arbitrarily close to perfect. This brings us to the more general version of randomized transmissions.

Repeating the Protocol for Stronger Security
Let us recall that we can split the message m into any number of parts $s_1, s_2, \ldots, s_t$. Suppose that we transmit each of these over a randomly selected path with the probability prescribed by p^*. Assuming stochastic independence between the transmissions and a mobile adversary which can attack a different set A in each round, the probability of a confidentiality breach from (7.3) becomes

$$\Pr\left(\text{message is disclosed} \,\middle|\, \begin{array}{l} \text{random routing according to } p^*, \\ \text{and \textit{arbitrary} attack patterns} \\ \text{based on } \mathscr{A} \end{array}\right) \leq (1 - v)^t$$

$$\to 0 \text{ as } t \to \infty.$$

The price paid for this increase in confidentiality is t-fold network traffic since each part s_i is equally long as m. Hence, nothing is free here (as is barely anything anywhere).

Using More Paths
The message parts are not required to traverse the network sequentially one by one, but can flow concurrently in groups of two or more if we use multiple paths simultaneously.

There is no conceptual change to the procedure except for replacement of the list L in the above tableau with a longer list of subsets of paths. If we use up to t paths concurrently, the number of subsets is of the order $O(|L|^t)$, and for practical feasibility, we may confine ourselves to an enumeration of only a subset of these to include in the tableau. If the value v is unacceptably low, then we can remove some path bundles and add new ones to recompute v accordingly. The indicator variables u_{ij} then depend on how many paths are used and how many of them are accordingly intercepted by the attacker, but the semantics and question of where to put a 0 and where to put a 1 in the tableau remains the same, and as with (7.1), only according to the new setting.

Incorporating TVA Information

Naturally, we may ask whether it is possible to use more of the information given by a TVA, for example:

- Can we account for the difference in robustness of individual nodes in the transmission?
- Can we include information about the difficulty of attacking node v_i, as opposed to the challenge of hijacking another node v_j?
- Can we include information about the necessary skill level for an attacker to mount attacks on node v_i, although it may require another set of skills or resources to attack node v_j? Can we make use of this information with the security quantification method if it is available from a TVA?

The answer to these questions is affirmative, and again, the change in the conceptual framework above is merely the difference in setting the indicator u_{ij}: nothing forbids us from using values in the entire range $[0, 1]$ rather than only the extremes 0 or 1 when filling the tableau. For example, a path π_i is selected, and the attacker attacks the set $A_j \subseteq V$ of nodes on which the particular node $v \in V(\pi_i) \cap A_j$ is situated. Now, let us assume that we know the node v is hardened in some manner which allows us to (heuristically) quantify the probability of the intruder sneaking in as the value

$$u_{ij} = 1 - \Pr(\text{intruder hijacks the node } v \mid \text{given the "hardness" of } v).$$

A value v computed from this matrix inherits the interpretation as a probability, and all the previous results, in particular (7.3), remain valid under this setting.

Matters change slightly if numbers other than probabilities are applied, for example, CVSS scores. The computation as such, however, remains unchanged. Let us re-examine our previous example under the different consideration of TVA to measure the difficulty of hacking a node:

Example 7.4 Consider the network illustrated in Fig. 7.3, although this time with the additional information of CVSS scores for all nodes. Let these scores be:

Node	1	4	5	6	7
CVSS	7.4	8.2	9.1	8.0	7.6

As before, we need a method to set payoffs in the eavesdropping game if Alice chooses a certain path and the attacker strikes a certain node. For simplicity, let us adopt only Alice's perspective and directly assign the CVSS score as a payoff. In this way, (7.2) reads as Alice seeking the path which has the maximum score, while the attacker conversely strikes the path which has the minimum score, i.e., the easiest to intercept. If the attacker misses the transmission by attacking a node which Alice does not use to relay her data, we note a full score of 10, which is the maximum on the CVSS scale.

The resulting game model now carries the CVSS scores for the nodes wherever a zero was previously and a 10 for all the 1s:

	Attacked node				
↓path	1	4	5	6	7
π_1	10	8.2	10	10	10
π_2	7.4	8.2	10	10	10
π_3	7.4	10	9.1	10	10
π_4	10	10	10	8.0	7.6

Computing the value v, we obtain $v \approx 9.26$, and the best strategies for using the paths have the following probabilities:

$$
\begin{array}{|c|c|c|c|c|}
\hline
\text{path} & \pi_1 & \pi_2 & \pi_3 & \pi_4 \\
\hline
\text{probability} & 0.40941 & 0 & 0.28328 & 0.30731 \\
\hline
\end{array}
\tag{7.4}
$$

Similarly, the attacker's best behavior would be striking the nodes according to their individual vulnerabilities, arriving at the optimum

Node	1	4	5	6	7
Probability	0.287	0.408	0	0	0.305

It is, however, important to remark that the value of v hereby *loses* its interpretation as a probability; it is now merely the *best achievable CVSS score*. Whether this is a useful quantification of security for the given context is for the security officer to decide.

The optimum values for p^* and q^* shown in the above table, however, *do remain* the best routing strategies for maximizing the security measure v.

7.2.1.1   Computing the Probability of Eavesdropping

The practical application of the method as outlined is a matter of computing Nash equilibria in matrix games, which entails solving a linear optimization problem.

Specifically, if $U \in \mathbb{R}^{n \times m}$ is an $(n \times m)$-matrix representing the tableau which models the transmission as a game (abstractly allowing the sender to choose between n configurations for the transmission and the adversary having $m = |\mathscr{A}|$ attacks to possibly mount), then, the value v and optimum p^* is found by solving the following (primal) Linear Program (LP) with $p = (p_1, p_2, \ldots, p_n)$:

$$(P) \quad \min v \quad \text{subject to} \quad \left(\frac{-A^T \mid 1}{1 \mid 0} \right) \cdot \begin{pmatrix} p_1 \\ \vdots \\ p_n \\ v \end{pmatrix} \begin{matrix} \geq \\ \vdots \\ \geq \\ = \end{matrix} \begin{pmatrix} 0 \\ \vdots \\ 0 \\ 1 \end{pmatrix},$$

and $p_i \geq 0$ for all $i = 1, \ldots, n$.

The optimal choice rule over the attacker's set $\mathscr{A}$ is similarly computed by solving the dual program to (P), which functions according to standard techniques. As with the solution to (P), computing the adversary's best attack pattern q^* is not unique, and therefore accepting the modal (maximum) among the probabilities as an indication of where to expect an attack is generally *not reliable* since the dual program can have a multitude of optima, all of which may provide different indications of where an attack is most likely (indeed, the set of optima is either singleton or convex and hence infinite, although it is never empty).

However, the non-uniqueness of the optima in the primal or dual program above has no problematic effect for the sender since the value v which determines the bound is *the same* for all optima. That is, the sender can select *any* optimal solution of (P) to govern its transmission and retain the same bound (7.3) for eavesdropping. Taking the attacker's optimum as a pointer to weak spots in the network is ambiguous and hence problematic.

7.2.2   Quantifying the Probability for a DoS

As with confidentiality, the semantics of the indicators u_{ij} can be set to account for events of transmission failure. In that case, the same bounds as above apply yet relate to the probabilities of a DoS, while all computational matters remain unchanged. As a further generalization, we can replace the XOR-decomposition of messages with polynomial secret sharing. This has the appeal of being secure against passive and active attacks [11, 12] and also being closely related to the Reed-Solomon code [13], which permits error correction to some extent. This brings us to a simple change in computing the message parts from the humble XOR to using polynomial secret sharing, knowing that error correction algorithms such as Welch-Berlekamp are available for recovery from packet loss. One protocol which embodies this idea is found in [14].

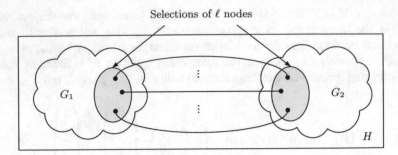

Fig. 7.5 Joining two ℓ-connected networks G_1, G_2 into an ℓ-connected network H

The change in modeling then only relates to correctly setting the indicators according to whether a message can be either reconstructed or disclosed (the two events need distinct assessment generally, and in particular, whether polynomial secret sharing is used).

Constructing a Network Topology with High Connectivity

Whitney's Theorem 7.1 provides a theoretical yet impractical test for the existence of node-disjoint paths, but it lends itself to an efficient method of constructing networks with prescribed connectivity from the ground up. Indeed, the smallest network (in terms of the number of nodes and edges) to be ℓ-connected is the complete graph with $\ell + 1$ nodes. These are the smallest initial networks which accomplish an "inductive start".

Given two networks $G_1 = (V_1, E_1), G_2 = (V_2, E_2)$, each with vertex connectivity ℓ, we can ("induction step") select ℓ distinct nodes $u_1, \ldots, u_\ell \in V_1$ and $v_1, \ldots, v_\ell \in V_2$ and connect them pairwise with a new set E of links between G_1 and G_2 as $E = \{\{u_1, v_1\}, \{u_2, v_2\}, \ldots, \{u_\ell, v_\ell\}\}$, into a new graph $H = (V_1 \cup V_2, E_1 \cup E_2 \cup E)$.

Applying Whitney's theorem 7.1, it is not difficult to prove (see [2, Chp.2]) that the resulting graph H is again ℓ-vertex connected and thus has the same number ℓ of node-disjoint paths. Figure 7.5 illustrates the connection between networks. The selected nodes technically correspond to entry and exit gateways, and the finding from Whitney's theorem is merely the observation that joining a network with $\ell \geq 1$ connections or bridges will eventually lead to stronger connectivity which not only increases availability but is also useful in gaining stronger confidentiality.

7.2.3 Quantifying Multiple Security Goals

If security is viewed as the joint goal of confidentiality and availability (the latter understood not only as the delivery of a message but also as unmodified delivery, i.e., including the goal of integrity), it is natural to ask whether the game-theory method of quantifying security extends to multiple indicators. And indeed, it does, if we model each security goal according to its own tableau (and game) and aim for multi-criteria optimization toward a Pareto-optimal bound for several goals. Let us first look at how security is quantifiable from a joint viewpoint by applying different models for confidentiality and availability.

Definition 7.1 A protocol is called δ-*reliable* if the delivery of a message succeeds with a probability of at least $1 - \delta$. A protocol is ε-*private* if the two protocol transcripts for messages M_1 and M_2 acquired by the adversary have distributions which differ at most by 2ε in the 1-norm. Formally, we need to show that if $T(M_1, r), T(M_2, r)$ denote the adversary's *transcript* of the entire protocol execution for the delivery of two messages M_1 and M_2, where r represents the random coin-flips, then for any r, we require $\sum_C |\Pr(T(M_1, r) = C) - \Pr(T(M_2, r) = C)| \leq 2\varepsilon$ (note that we assume the same coin-flips in the transmission of one or the other message).

With this definition, we have the following formal results from the above modeling and application of game theory:

Theorem 7.2 *The saddle point value v from (7.2) is a measure of privacy and reliability in the following sense:*

1. *If Alice and Bob set up their model matrix with entries u_{ij} such that*

$$u_{ij} = \begin{cases} 1, & \textit{if the message is delivered successfully;} \\ 0, & \textit{otherwise.} \end{cases}$$

   for every strategy combination of honest party and adversary, then the protocol is $(1 - v)$-reliable.
2. *If Alice and Bob set up their model matrix with entries u_{ij} such that*

$$u_{ij} = \begin{cases} 1, & \textit{if the adversary learns nothing about the secret content;} \\ 0, & \textit{otherwise.} \end{cases}$$

   for every strategy combination of honest party and adversary, then the protocol is 2ε-private, where $\varepsilon = 1 - v$.

A corollary of this result formalizes a converse to the sufficiency of the security measure v:

Theorem 7.3 *Let Alice and Bob model their communication as a game with a binary matrix $u_{ij} \in \{0, 1\}$, where $u_{ij} = 1$ if and only if a message can be delivered securely by choosing the i-th pure strategy and the adversary uses its j-th pure strategy for attacking. Let v be the saddle point value of the game according to (7.2), then $v \in [0, 1]$, and*

1. *for any $\varepsilon > 0$, if $v > 0$, then a protocol exists so that Alice and Bob can communicate with an eavesdropping probability of at most ε.*
2. *if $v = 0$, then the probability of the message being extracted by the adversary is 1.*

3. *If we consider Alice a random source which emits messages M with Shannon-entropy $H(M)$, then the information h which leaks to an eavesdropper is bounded as*

$$h \leq (1 - v) \cdot H(M)$$

Therefore, knowing that the saddle point value v computed from respectively established models provides us with security bounds in each respect, let us recall that the optimal strategies in applying the bounds are nonetheless individually distinct for privacy and availability. For a joint consideration, we need a randomized routing rule p^* which achieves one bound for confidentiality/privacy and another bound for availability. It is not surprising that a trade-off exists between these two goals if we jointly optimize them. To this end, we replace the scalar bound v in (7.3), which applies to a single goal, with a vector of similar bounds $v_1, v_2, \ldots$ for application to multiple security goals such as 1: confidentiality, 2: availability, and so on.

Toward an axiomatic approach, we can set (7.3) with the requirement for a multitude of bounds $(v_1, v_2, \ldots)$ for several security goals and request for the existence and algorithms to compute them. This is called Multi-Goal Security Strategy (MGSS), informally defined as a randomized choice rule p^* over all transmission configurations of the sender, together with a set of values $v_1, v_2, \ldots \in \mathbb{R}$, one per security goal, achieving two requirements:

Assurance: It should give a guaranteed, and with respect to the model, provable level of quality for the services it refers to. Formally, we require that Pr(breach on security goal i) $\leq v_i$ whenever the sender implements the optimal transmission according to p^*,

Efficiency: Any other randomized transmission rule $p' \neq p^*$ would invalidate at least one of the bounds, i.e., whenever the transmission is executed according to p' instead of p, we have Pr(breach on security goal j) $> v_j$ for at least one goal j (depending on p'). In this case, we also have a concrete attack strategy to exceed the security bound v_j strictly.

The second requirement equivalently states that we cannot uniformly improve protection upon implementing a randomized transmission differently from p^*. Therefore, the randomized transmission rule p^*, as in the previous case of single-path transmission, is an OLA. Under this OLA, the bounds implied by the assurance property as stated above hold, so that they may be taken as an SLA regarding the QoS in multiple dimensions.

Again, there is no conceptual requirement for multiple objectives to be only concerned about security. Indeed, especially in quantum networks, the interest may be on quality of service which *includes* but is *not limited* to security. Let us demonstrate what MGSS means if we consider a non-security requirement such as bandwidth.

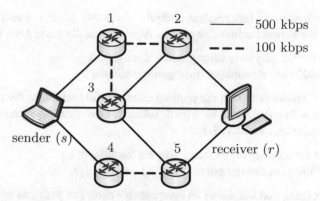

Fig. 7.6 Network with a vulnerable location between Alice and Bob

Example 7.5 (Bandwidth-Security Trade-Off in a Quantum Network) The example is taken from [15], where additional details may be investigated. We briefly summarize here, starting from the simple network in Fig. 7.6, in which all links run QKD but at different bandwidths imposed by the individually differing routing hardware.

As before, but somewhat more generally, let us assume that Alice employs two paths for each transmission, providing a respective indicator of security breach if and only if *both* paths are intercepted. Accordingly, the effective bandwidth averages between the selected paths depending on the traffic volume over each path. We simplify matters by not considering any concurrent traffic from other communicating peers, which could in a real world scenario randomly reduce the available bandwidth. Let us suppose that the bandwidths indicated in Fig. 7.6 are guaranteed to everyone who uses a link.

We can then set up a matrix of indicators and another matrix of bandwidths according to the selected paths. Refraining from a tedious enumeration, we obtain ten options for pairs of two paths for transmission from the sender to the receiver, here denoted π with a superscript 2 to indicate that this is *two* paths. Likewise, we have ten size-2 subsets of nodes to attack, also enumerated as attack actions $A_1, \ldots, A_{10}$ for the adversary. For compactness, let us compile both into a vector-valued matrix, modeling the multi-goal transmission game as:

	A_1	A_2	A_3	A_4	$\cdots$	A_9	A_{10}
π_1^2	(0,300)	(1,300)	(1,300)	(1,300)	$\cdots$	(1,300)	(0,300)
π_2^2	(0,550)	(1,550)	(1,550)	(1,550)	$\cdots$	(1,550)	(1,550)
π_3^2	(0,100)	(0,100)	(1,100)	(1,100)	$\cdots$	(1,100)	(0,100)
π_4^2	(1,100)	(0,100)	(0,100)	(1,100)	$\cdots$	(1,100)	(0,100)
$\vdots$	$\vdots$	$\vdots$	$\vdots$	$\vdots$	$\ddots$	$\vdots$	$\vdots$
π_9^2	(1,300)	(1,300)	(0,300)	(1,300)	$\cdots$	(1,300)	(0,300)
π_{10}^2	(1,550)	(1,550)	(0,550)	(1,550)	$\cdots$	(1,550)	(1,550)

Proceeding with an independent analysis of the two goals (1) secrecy and (2) bandwidth, the optimal achievable results, regardless of the other goal, are

- $v = 2/3$ for security only (disregarding throughput),
- and $v = 400$ kbps throughput (disregarding security).

The point of MGSS is to find the optimal balance while retaining the assurance of performance with reference to each goal. Taking the two goals as equally important, an optimal balance, i.e., MGSS, is:

- $v_1 = 1/2$ for security only (accounting for throughput),
- and $v = 550$ kbps throughput (accounting for security),

both aspects being conditional on an equiprobable choice of path sets $\pi_2^2 = \{s—3—r, s—1—2—r\}$ and $\pi_{10}^2 = \{s—4—5—r, s—3—r\}$. The increase in bandwidth here is obtained at the cost of weaker privacy (lower v_1), and conversely, if the security is increased by repeating the transmission t times, the bandwidth is reduced by a factor of t. This is, intuitively, the mechanism behind this particular trade-off, but it proves to be an inherent balance which is unavoidable. "Optimality" here indicates the impossibility of gaining both greater security and more bandwidth simultaneously; it is simply unobtainable.

It can be shown (see [15]) that MGSS always exist and that computing them can be reduced to a sequence of LPs, as outlined in Sect. 7.2.1.1. The entire algorithm has been implemented in the HyRiM package [16] released under GNU Public License (GPL) and is available in the free language R [17] for statistical computing.

MGSS treats all security goals on equal grounds and intrinsically imposes no order of preference between them. Whether we rate confidentiality higher than availability is an aspect which can be included by assigning importance weights to the goals (these are input directly into the computational routines of the HyRiM software as parameters for computing MGSS). An alternative method of computing optimal transmission routes while applying a strict order of preference to goals is by computing v, p^* and q^* as an equilibrium over a lexicographic order. As an algorithm, this entails solving a sequence of suitably crafted LPs, one for each security goal, but it is computationally more expensive since it requires finding all optima for each LP. We refer to [18] for details and the freely available code.

7.3   Weaponizing the Detection of Eavesdropping

An implicit condition for the validity of (7.3) or its generalizations is the strength of the routing. What happens if by coincidence the selected paths are not available or due to circumstances are beyond the sender's control? More specifically, is secure transmission also possible if the attacker tampers with the routing? The short answer is "yes", as long as the routing has not been "disrupted too much".

Indeed, the eavesdropping facility of QKD can be transformed into an attack device, allowing the adversary to temporarily block paths by passively listening in on them. The intention here is not to read any information—this is precluded by the security proofs of QKD—but the goal is rather a humble DoS on the line.

Suppose that such passive eavesdropping attacks are mounted to force the routing mechanisms to send all flows over a single or a few nodes which the adversary has under its control. We call this *indirect eavesdropping* because listening is executed on one line to reveal the information at another location in the network, possibly very distant from the line under attack. It may not even require listening on the quantum line, as simply overloading a network link may already force the nodes to seek alternative routes which are less crowded. This can also lead to re-routing over nodes which are under the adversary's control (see Fig. 7.7 for an illustration).

Referring to the literature [19] for details, let us confine ourselves to outlining the basic mechanism for retaining security against routing distortions: in summary, under random distortions, we can consider routing to be a Markov chain which describes the random walk of a packet over a network graph. This random walk is governed by a ($|V| \times |V|$)-matrix P of probabilities which informs how likely a packet will be routed from node v_i to node v_j in the network and models the now probabilistic routing behavior in contrast to the formerly presumed guaranteed transmission paths selected by the sender.

Then, based on matrix P to describe a packet's travel from s to d as an instance trajectory of a Markov chain, we can find the conditions for a packet to *bypass* a given set of nodes under the adversary's control. Assuming that the message is recovered from the entirety $s_1, \ldots, s_t$ of packets as $m = s_1 \oplus s_2 \oplus \cdots \oplus s_t$, the bypassing packet which escapes the adversary operates as a one-time pad on message m, thereby preventing it from disclosure (only). Note that this does not apply to more sophisticated encodings with error correction, as discussed in Sect. 7.2.2, and that the reliability of packet delivery under random distortions in routing is another topic.

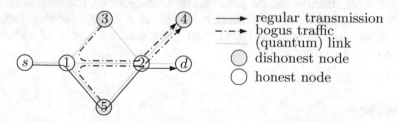

Fig. 7.7 Forced alteration of the path by overloading selected links with bogus traffic

Table 7.1 Correspondence of SLA and OLA to the output artefacts from a game-theoretic security modeling and analysis

Object	Game-theoretic counterpart
SLA	Saddle-point value or MGSS assurance values $v_1, v_2, \ldots$ of the multipath transmission multi-criteria game. These values may relate to QoS aspects of interest, such as confidentiality, bandwidth, etc., as in the Example 7.5
OLA	Optimal randomized routing strategy p^* (see (7.4) for an example) as computed from the Nash equilibrium for the defender in the transmission game

7.4   Summary

The security of quantum networks is unconditional if peers share direct links, but it is also quantifiably extensible to end-to-end security, with or without trusted relays. For an end-to-end QoS statement, a user can ask for the probability of eavesdropping, information leakage in terms of entropy, resilience to link failure to DoS attacks, bandwidth, and others. Trusted relays provide a simple approach that equates point-to-point with end-to-end security in all respects, but essentially, the chain will be as strong as its weakest link. This applies to all the aforementioned dimensions, because bandwidth will be determined by the slowest link, eavesdropping is easiest to achieve on the node with the weakest protection, link failure is the most likely at the least protected link, and so on. Exploiting the (in many instances given) redundancy in network connectivity, we can quantify security according to many relevant dimensions. From a security perspective, we can guarantee that the quantified values for the probability of eavesdropping and the likelihood of denial of service, bandwidth, and others are guaranteed within the anticipated set of attack scenarios. The results obtained from a game theoretic analysis of randomized multipath transmission then yields artifacts which are interpretable as SLA and OLA, as Table 7.1 details.

Both the saddle point and assurance values are systematically obtainable from the QKD network structure and the properties of its components. Information such as CVSS scores for network components, the resilience of nodes, bandwidth capacity of links, etc., may all be incorporated, leaving the numerical matters of computing MGSS to existing software [16].

References

1. Schartner, P., & Rass, S. (2010).   Quantum key distribution and denial-of-service: Using strengthened classical cryptography as a fallback option.   In *International Computer Symposium (ICS)* (pp. 131–136). IEEE.
2. Rass, S. (2009). *On Information-Theoretic Security: Contemporary Problems and Solutions.* PhD Thesis, Klagenfurt University, Institute of Applied Informatics.
3. Chartrand, G., & Zhang, P. (2005). *Introduction to graph theory.* Higher Education. Boston: McGraw-Hill.

4. Rass, S., Wiegele, A., & Schartner, P. (2010). Building a quantum network: How to optimize security and expenses. *Springer Journal of Network and Systems Management, 18*(3), 283–299. https://doi.org/10.1007/s10922-010-9162-0

5. Jajodia, S., & Noel, S. (2010). Topological vulnerability analysis. In S. Jajodia, P. Liu, V. Swarup, & C. Wang (Eds.), *Cyber situational awareness*. Advances in Information Security (Vol. 46, pp. 139–154). Boston, MA, USA: Springer. ISBN 978-1-4419-0139-2, 978-1-4419-0140-8. https://doi.org/10.1007/978-1-4419-0140-8_7.

6. Information Technology Laboratory. (2020). *Nvd - home*. https://nvd.nist.gov/

7. Houmb, S. H., & Franqueira, V. N. L. (2009). Estimating ToE risk level using CVSS. In *2009 International Conference on Availability, Reliability and Security* (pp. 718–725). IEEE Computer Society Press.

8. Rass, S., Rainer, B., & Schauer, S. (2013). On the practical feasibility of secure multipath communication. *International Journal of Advanced Computer Science and Applications, 4*(10), 99–108. https://doi.org/10.14569/IJACSA.2013.041016

9. Rass, S., Rainer, B., Vavti, M., & Schauer, S. (2013). A network modeling and analysis tool for perfectly secure communication. In *2013 Ieee 27Th International Conference on Advanced Information Networking and Applications (Aina)* (pp. 267–275). ISSN 1550-445X. https://doi.org/10.1109/AINA.2013.34

10. Rass, S., & Schartner, P. (2011). Information-{leakage} in {hybrid} {randomized} {protocols}. In J. Lopez & P. Samarati (Eds.), *Proceedings of the {International} {Conference} on {Security} and {Cryptography} ({SECRYPT})* (pp. 134–143). SciTePress – Science and Technology Publications. ISBN 978-989-8425-71-3.

11. Rabin, T., & Ben-Or, M. (1989). Verifiable secret sharing and multiparty protocols with honest majority. In *Proceedings of the Twenty-First Annual ACM Symposium on Theory of Computing* (pp. 73–85).

12. Tompa, M., & Woll, H. (1989). How to share a secret with cheaters. *Journal of Cryptology, 1*(3), 133–138. ISSN 0933-2790, 1432-1378. https://doi.org/10.1007/BF02252871

13. McEliece, R. J., & Sarwate, D. V. (1981). On sharing secrets and reed-solomon codes. *Communications of the ACM, 24*(9), 583–584.

14. Fitzi, M., Franklin, M., Garay, J., & Vardhan, S. H. (2007). Towards optimal and efficient perfectly secure message transmission. In *Theory of Cryptography Conference* (pp. 311–322).

15. Rass, S. (2013). On Game-Theoretic Network Security Provisioning. *Springer Journal of Network and Systems Management, 21*(1), 47–64. https://doi.org/10.1007/s10922-012-9229-1

16. Rass, S., König, S., & Alshawish, A. (2020). *R package 'hyrim': Multicriteria risk management using zero-sum games with vector-valued payoffs that are probability distributions, version 2.0.0.* https://CRAN.R-project.org/package=HyRiM.

17. R Core Team. (2020). *R: A Language and Environment for Statistical Computing.* R Foundation for Statistical Computing, Vienna. https://www.R-project.org/

18. Rass, S., Wiegele, A., & König, S. (2020). Security games over lexicographic orders. In Q. Zhu, J. S. Baras, R. Poovendran, & J. Chen (Eds.), *Decision and game theory for security*. Lecture Notes in Computer Science (Vol. 12513, pp. 422–441). Cham: Springer International Publishing. ISBN 978-3-030-64792-6, 978-3-030-64793-3. https://doi.org/10.1007/978-3-030-64793-3_23

19. Rass, S., & König, S. (2012). Turning quantum cryptography against itself: How to avoid indirect eavesdropping in quantum networks by passive and active adversaries. *International Journal on Advances in Systems and Measurements, 5*(1), 22–33.

Chapter 8
Modern Trends in Quantum Key Distribution Networks

The development of electronic and optical components has led to greater interest in the application of QKD solutions in everyday life. In this section, we briefly describe the currently attractive trends and approaches for future research in the field of QKD technologies.

8.1 QKD in 5G Networks

With the faster deployment of 5G telecommunications systems, security and user privacy is becoming increasingly vital. Networks created for superior performance and to provide support for the purposes of autonomous vehicles, remote surgery, autopilot functions, remote vehicle management, smart-grid systems and other functions which involve life and death situations primarily depend on reliable end-to-end secure network connections. The vision behind 5G lies in providing exceptionally high data rates with minimum latency, manifold increments in base station capacity and the overall density required to significantly improve the QoS of future 5G network applications [1].

Although multiple structural models for organizing telecommunications networks are available, the Cisco model is often used as a reference [2]. It consists of three layers:

- The core layer provides the backbone with uninterrupted connectivity across the network. Its main aim is to provide fast, reliable service at national and international levels.
- The distribution layer is a multipurpose layer which aggregates and forwards the traffic received from the access layer before it is forwarded to the core layer for routing to its final destination. Together with the core layer, it constitutes the transport network for providing functionality for the transport, multiplexing,

© Springer Nature Switzerland AG 2022
M. Mehic et al., *Quantum Key Distribution Networks*,
https://doi.org/10.1007/978-3-031-06608-5_8

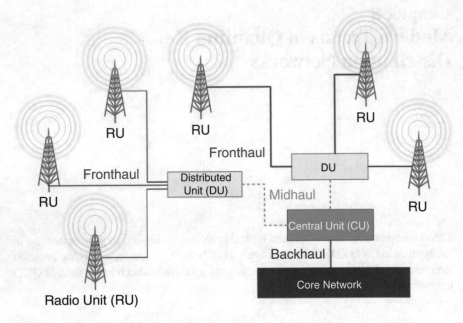

Fig. 8.1 5G Structure: fronthaul, midhaul and backhaul sections

switching, management and supervision of optical channels used for the transmission of client signals.

• The access layer connects clients with the network. It is distinguished by short distances as a result of strict limits to costs. In the previous generation of mobile networks, the access layer was generally built on copper to bring broadband to fixed subscribers and to connect mobile antenna sites to the remainder of the network. However, operators are increasingly switching their access networks to optical fiber to provide the faster services which characterize 5G networks.

Figure 8.1 illustrates the three typically considered systematic sections of the 5G access layer: fronthaul, midhaul and backhaul.

Radio Access Network (RAN) is often referred to as the *fronthaul* network segment which oversees communication with the User Equipment (UE). In previous mobile network generations, the base station was contained in a single physical box. This configuration led to a distributively represented structure. The diagram in Fig. 8.2a shows that it was usually connected to an antenna with a coaxial cable that was unable to support modern high-speed requirements, resulting in significant losses.

Advances in the development of electronic components allowed multiple radio communication blocks to be merged into a single unit. It then became possible for a single unit to operate with multiple RF antennas. RF amplifiers were deployed on the antenna to reduce attenuation in the coaxial cable connecting the antenna, while radio processing was performed away from the antenna array, as illustrated

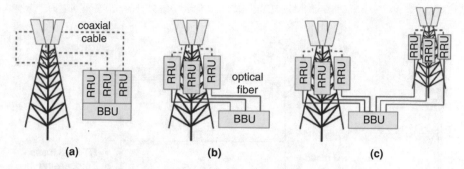

Fig. 8.2 RAN evolution: (**a**) Distributed RAN; (**b**) Migration to C-RAN; (**c**) C-RAN structure

in Fig. 8.2b. This type of structure led to the concept of centralized (C-RAN) architecture, in which the base station consists of a radio digital processing unit called the Digital Unit (DU) or Base Band Unit (BBU), which serves multiple RF operation units called Remote Radio Head (RRH) or Remote Radio Unit (RRU), or simply Radio Unit (RU), as shown in Fig. 8.2c [3].

The BBU is used to process baseband signals (signals before modulation) through the physical interface to the core network. The RRU consists of several RF processing units and cell site antennas. To ensure high communication speeds between the BBU units, which are located in the secure Central Office (CO) and external RRU, optical connectivity is required. This, however, is limited to tens of kilometers to satisfy the timing requirements of the radio technology. The Hybrid Automatic Retransmit reQuest (HARQ) protocol is primarily used as a retransmission mechanism between the UE and RRU and foresees less then 3 ms for DU processing. To reduce the delay caused by fiber propagation and to satisfy HARQ requirements, the fronthaul link which connects the RRU and BBU should be implemented over a distance of about 20 km, depending on the RAN implementation (Fig. 8.3) [3–5].

Connectivity for end users in 5G networks is characterized by heterogeneous access (HetNets) to a large number of smaller network nodes with high bandwidth and minimal delay, categorized by both energy and computational sensitivity. 3GPP has developed 5G New Radio (NR), a new Radio Access Technology (RAT) for the 5G mobile network, forming the foundation for the next generation of mobile networks. It is designed to attain the same wireless broadband performance as fiber cabling while having a substantially greater spectral efficiency. 5G NR aims to efficiently connect Internet of Things (IoT) devices and offers new types of mission-critical services by lowering latency and increasing reliability and security [6–8]. 5G NR considers SA (stand-alone) and NSA (non-standalone) operation based on ongoing technical work and includes two frequency ranges:

- Sub-6 GHz (less then 6 GHz)
- mmWave (above 24 GHz)

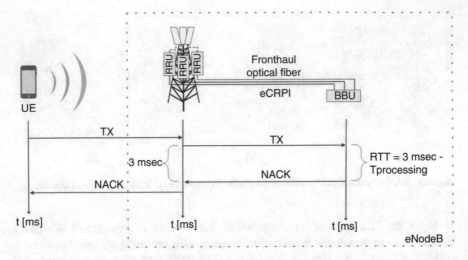

Fig. 8.3 Latency link budget of MAC-layer RTT in C-RAN

According to 3GPP, the *midhaul* refers to a group of links which connects the fronthaul and backhaul segments [9]. The OPEN RAN 5G evolution has enabled modification of the structure of BBU and RRH architecture defined for 4G networks by separating the BBU into a Distributed Unit (DU) and a Central Unit (CU). Some of the traditional BBU entity functionalities such as PHY, MAC and RLC sublayers have been retained in the DU, while other sublayers have been relocated to the CU. The midhaul thus connects the DU and CU units. The 5G network evolution has also led to the integration of the RRH and antenna system into a single unit, now called the Active Antenna Unit (AAU) [3].

This type of arrangement requires the fronthaul to perform fast communication and satisfy the comprehensive stringent QoS requirements of 5G networks. It also means that the fronthaul link to RU units is the only connection which extends beyond the physical facilities of the telecommunications operator, and therefore, its security should be specifically addressed. The most common solution for fronthaul links is Enhanced Common Public Radio Interface (eCPRI), which allows the integration of MACsec and IPsec solutions[1] [11, 12]. Because they can be implemented using symmetric keys (Sect. 2.4.4), the application of QKD is indicated. The application of additional cryptographic techniques would certainly

---

[1] The term *post-quantum cryptography* [10] is a special subfield of cryptography, comprising primitives, especially classic non-quantum schemes, which are potentially resilient to attacks from quantum computers. Such schemes, however, are mainly built from classic components and do not necessarily rely on any quantum technology themselves. For example, McElice public-key encryption, based on linear coding problems, is a prominent example of a post-quantum scheme. The assumptions that QKD makes to assure security likewise extend beyond attackers possessing quantum computers and even allow for infinite computational power (embodied in the definitions of ITS security), regardless of whether a quantum computer is present.

lead to additional overhead and reduce the sensitive performance of the fronthaul link, but given that these links are limited to a distance of several tens of kilometers, the application of a QKD system is highly attractive. Since symmetric encryption is faster than an asymmetric solution [13], the application of QKD has potential as long as sufficient cryptographic keys are available. The management and orchestration of key storage is therefore highlighted as a critical component in further research of QKD systems.

The integration of QKD with 5G mobile fronthaul network segments has been the subject of several studies [14, 15]. The studies evaluated the performance of QKD systems under the constraints of distances imposed by tight latency requirements. The performance of the fronthaul system was also investigated through a combination of quantum and public channels over a single optical fiber. Raman scattering decreased the performance of the quantum channel, although a greater spectral distance between the quantum and the public channels could have been achieved with a split fiber or a separate dark fiber to realize the quantum channel.

Efforts are underway to apply a CV-QKD system which increases QKD key generation speeds [16]. Unlike DV-QKD systems which rely on slow single-photon detectors, CV-QKD can be implemented using balanced homodyne detectors to achieve higher key generation speeds (Sect. 1.2.3).

To support working with a large number of remote RU units, the 5G fronthaul implements WDM techniques which enable increased bandwidth. This approach

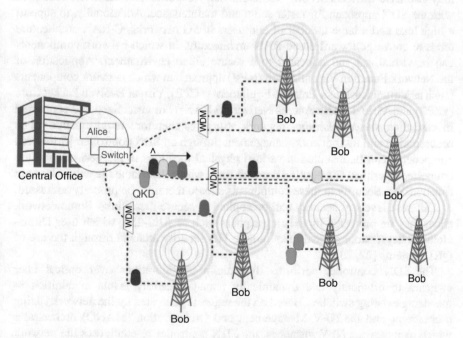

Fig. 8.4 Application of QKD in a 5G mobile fronthaul network

can also be used to integrate CV-QKD solutions with the classic channel [17, 18]. Figure 8.4 illustrates how a single QKD transmitter device can be connected for time-sharing with multiple QKD receiver systems that use spatial switching and spectral multiplexing. However, this type of implementation depends on strict time synchronization and the ability to store a large number of keys in case keys cannot be established because of poor time synchronization or other obstacles.

The *backhaul* network segment involves network components which provide communication between the base stations and the network core. The backhaul can be implemented using various communications technologies, such as wireless (satellite or microwave links) or dedicated optical pairs [19, 20]. The different applied technologies reflect the difficulty of guaranteeing the security of the particular network components. However, when the C-RAN architecture is applied, the BBU is installed in a secure central location. The application of additional IPsec protocols which impose an additional overhead can be considered redundant in the backhaul segment. Instead, security in this part of the network can be attained through SDN network slicing, which allows transformation into a virtualized simplified environment [21]. In practice, network slicing is the implementation of multiple separate virtualized networks over the same physical architecture. Although these virtual networks may have different network QoS requirements (e.g., networks which are used for communication between autonomous vehicles, which have very low latency but also rather low bandwidth, or networks for high-definition video streaming, which requires high bandwidth and low latency), they may also have different security requirements. Network slices are often combined with the SDN approach for faster setup and maintenance. Additionally, to support a high load and a large number of end-users in 5G networks, C-RAN architecture tends to converge toward cloud-RAN architecture, in which network components can be virtualized and relocated to a secure cloud environment. Application of the Network Function Virtualization (NFV) approach in which network components (such as Virtual Customer Premise Equipment (vCPE), Virtual Evolved Packet Core (vEPC) and Virtual Radio Access Network (vRAN)) are virtualized and connected in cascading *service-chaining* network structures therefore requires appropriate orchestration and fast network management through an SDN controller. Virtualized components can be installed at various physical locations in the network and are connected to critical aggregating network links which determine the network's core functionality. Security in these components should therefore be properly addressed.

Different levels of security can be applied to secure these links. Some network slices may be protected by existing ciphers such as AES-256, which uses Diffie–Hellman keys. More critical links require ITS security attained through the use of QKD systems [22, 23].

The SDN controller performs dynamic re-configurations over optical fiber switches to efficiently and dynamically manage security levels in addition to managing existing switches. Based on the requests generated by the network slicing orchestrator and the NFV Management and Orchestration (MANO) orchestrator which manages the NFV instances, the SDN controller re-configures the network to dynamically adjust the network slice to specific needs. Various protocols such as OpenFlow and NetConf can be applied for this purpose [24]. Since the SDN controller contains detailed information about network states, it can adapt network flows to different requirements, support multipath communication and provide redundancy and additional security [25]. Here it is important to note that the SDN controller and the KMS node are not required to be physically implemented within the same unit and therefore need clear communication which provide effective management [26]. In addition to the EU H2020 OPENQKD project, which explores the application of QKD technology through practical use-cases, the NATO SPS G5894 Quantum Cybersecurity in 5G Networks (QUANTUM5) project is especially interesting in this area.[2] The project explores the practical application of QKD technology in 5G networks and is developing the first QKD 5G simulator.

8.2 Measurement-Device Independent QKD

In considering the security of a QKD link, the detector, or more precisely, a measurement unit, forms a critical component of any practical implementation. A solution such as Measurement Device Independent Quantum Key Distribution (MDI-QKD) provides a method of circumventing numerous security vulnerabilities [27]. It also allows limitations in the distances of the QKD system to be overcome [28].

MDI-QKD entails that Bob sends additional photons instead of detecting them. The measurement procedure is performed on Charlie, which is an intermediate relay node located between Alice and Bob, as illustrated in Fig. 8.5. Since there are no detectors installed on the Bob and Alice nodes, the detector side channel attacks do not compromise protocol security. Although an attacker may have complete control over the Charlie node, Alice and Bob can still establish a secret key if they are provided with the results of the measurement. These results are announced over the public channel but do not reveal details of the key itself. The Charlie node can thus also be seen as an *untrusted relay*. MDI-QKD includes several protocol steps:

- Alice and Bob prepare phase randomized weak coherent pulses.
- Alice and Bob use a polarization modulator to prepare pulses in one of the four polarization states by randomly selecting polarization (as in the case of BB84 described in Sect. 1.2.1).
- To reduce Eve's attempt to perform the photon-number-splitting attack, Alice and Bob can use an intensity modulator to modify the amplitude of the pulse and generate either a signal or a decoy.[3]

---

[2] More details available at www.quantum5.eu.

[3] The decoy technique is based on the use of an additional photon source to generate weak coherent laser pulses with a different photon number distribution. It has a higher average photon number than the original source but does not differ from the original signal pulses in other parameter states (i.e., wavelength). When these pulses are sent between the signals generated by the original generator, Eve is unable to distinguish the original signals and treats them all equally. However, after Bob

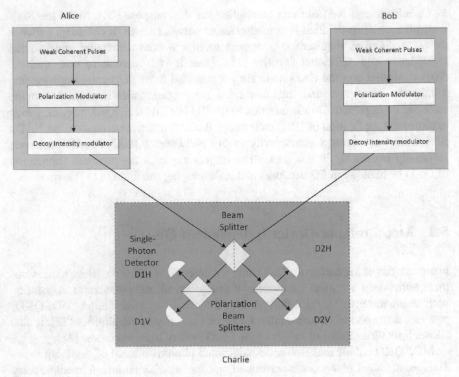

Fig. 8.5 Schematic diagram of MDI-QKD approach. Charlie's untrusted relay can be seen as "black box"

- Alice and Bob send prepared pulses to the beam splitter installed at Charlie's node. Each output port of the beam splitter is connected to the polarization beam splitter which transforms pulses into horizontal (H) or vertical (V) polarization states.
- The outputs of each polarization beam splitter are connected to two SPDs which detect the resulting photons.
- The results of detection are publicly announced.

As with the BB84 protocol, measurements are successful only in 50% of cases, more precisely, when photons are detected on SPDs:

- D_{1H} and D_{2V} or D_{1V} and D_{2H}, resulting in state $|\psi-\rangle$
- D_{1H} and D_{1V} or D_{2H} and D_{2V}, resulting in state $|\psi+\rangle$.

Other scenarios in which both photons are detected on the same detector cannot be applied because they reveal the polarization of the dispatched photons. In addition, situations where clicks are detected on photons D_{1H} and D_{2H} or D_{1V} and

receives the signals, Alice is able to inform him what the original signals were to allow easier detection of eavesdropping activities.

D_{2H} are ignored because of the Hong-Ou-Mandel (HOM) effect, which specifies that if two identically polarized photons are directed to the 50:50 beam splitter, they will always end up at the same output [27].

The obtained results are publicly announced, and Alice and Bob keep only those bits which result in successful measurement: specifically, those cases where they used identical bits with a rectilinear basis to encode their photons or cases where they used different bits applying diagonal basis. The obtained measurement results refer only to the parity values of the sent bits but not to the bits used to form the parity. The information obtained can therefore be used to generate secret keys between Alice and Bob [29].

Broadly speaking, MDI-QKD, or more generally Device-Independent Quantum Key Distribution (DIQKD), aims to relax assumptions on the quantum devices and considers the quantum components which Alice and Bob bring into the protocol as black boxes with classic inputs and outputs. Many security proofs of QKD to some extent rely on the following assumptions:

- Device reliability: this may concern the perfectness of channels, sources or detectors. All three assumptions are subject to exploitation in attacks, such as single photon sources nonetheless emitting multiple photons (beam splitting attacks) or Eve replacing the noisy channel with another channel that contains more or less noise, for example, as discussed in [30].

  Generally, the devices are intended to run continuously and endlessly, and attacks which seek to temporarily or permanently disable a device are not necessarily covered by the scope of QKD security proofs. As with trusted relays, organizational and other technical protective measures are responsible for reliability, although the term *reliability* needs a broader definition here: it also refers to the device precisely following expected specifications (again, corroborating the need for security certification and auditing, e.g., using Common Criteria (CC) or other methods). Shutting down a device is can be done with various methods and for various reasons, the most obvious being the use of the QKD eavesdropping detection mechanism to cause Alice and Bob think that an adversary is present and terminate the communication themselves or ultimately by running the key storage dry. An option for mitigation is fallback to classic encryption [31] but with the price of losing ITS security. This may, however, not be problematic, since it has been shown that optimal switching between "wasteful" yet perfectly secure OTP and "conservative" yet computationally secure symmetric encryption (e.g., AES) can retain the desired balance of security and efficiency at acceptable QoS levels [32].

- Reliability and robustness settings: the protocols typically assume that the configuration is under full and exclusive control of Alice or Bob, but attacks may seek to tamper with the device configuration [33] (also possibly using somewhat conventional techniques from (software) hacking, but, although reported, such trials have not been found in this survey).

- Confidentiality settings: judging from the number of attacks published concerning the methods of determining source or detector settings, this is one of the most critical areas where QKD devices can launch attacks. Quantum hacking is in many instances about determining or even manipulating the randomness of the QKD device, and the attempts are in no way limited to using quantum techniques only.

MDI-QKD seeks to relax one or more of the above assumptions, and in the best case, removes all of them [34, 35]. However, some of the more classic assumptions are often retained or still not considered by the relaxations which MDI-QKD targets [36]. These include the following additional prerequisites of many security proofs:

- Trustworthy classic components *within* the device, broken down into the following (at a minimum):

  - Trusted and perfect random number sources: the entire protocol depends vitally on unpredictable sources of randomness to control the emission of respectively polarized photons. Without even accessing any of the quantum components, an attacker can attempt to replace the source of randomness. A randomness substitution attack has already been described in the literature [37].
  - Trusted operating system and network stack: given the experimental nature of quantum devices, it appears natural to find them mostly running on open source operating systems since these offer the greatest flexibility in adapting the special needs of quantum devices. However, and regardless of the Operating System (OS) used, these are without exception subject to the full spectrum of classic hacking techniques which target the device software. Even with full drive encryption in place, side-channel attacks and cold boot attacks remain a danger to the "classic component" of QKD.
  - Protected administration: this concerns all interfaces and physical access to devices usually not included in the scope of security proofs. Securing administrative access requires the technical mechanisms of classic access control (e.g., role-based access using security tokens, biometric authorization, and in the worst case, a strong password), care at the organizational level, and attention to staff and workflow management for the provisioning of QKD devices. Respective security management standards are available (e.g., ISO27k) and can be instantiated for applications in quantum networks.

- Trust in components which QKD, according to the definition of its aims, is not able to cover, for example (at a minimum):

  - Authenticated channels,
  - Authenticity of devices generally (e.g., assurance that the device has not been assembled from parts which have already been tampered with by an adversary and implanted with hardware trojans, etc.); this assumption can partially be relaxed if Multipath Transmission (MPT) is in place (see Chap. 7).
  - The correctness of quantum physics itself.

Nevertheless, MDI-QKD has accomplished many improvements [35, 38–40] which enable the use of much less "ideal" quantum devices and have established the path toward further roll-out of quantum networks with reduced attack surfaces on the quantum side. The performance of MDI-QKD systems [41–43] has also improved with security aspects, for example, MDI-QKD may be able to increase the distance limits of QKD links compared to conventional QKD systems [44]. In this regard twin-field QKD is also notable [45], being similar to MDI-QKD but aiming to oversome the rate-distance limits of QKD.

8.3   Quantum Repeater

A quantum repeater [46–50] is a technology which bridges unlimited distances using several hops, as in trusted relays, but unlike a trusted relay, its aim is not to deliver a payload but to extend key distillation across several intermediate relays. The principles of conventional repeaters which amplify the signal do not apply here, since this amplification, under the no-cloning theorem, would destroy the quantum characteristics, and the damping of the fibre places a practical limit on the distance we can bridge. Quantum repeaters therefore use entanglement to transport information, as in quantum teleportation. In a nutshell, imagine a node which prepares an entangled pair of photons where one of the photons is sent to the first hop, being a quantum memory. Stored there, the qubit waits for the next segment to become ready, and the state can be entangled with the second hop, and so on, until the final hop, Bob's node obtains a state which is entangled with Alice's state, ideally at any desired distance. The practical challenge is to keep noise under control since the process of entanglement swapping becomes more and more difficult with each repetition. Clearly, the intermediate quantum memories are the vital components here, and a "trusted device assumption" may also apply to them. Currently, the concept is under intense research and many experimental demonstrations of all areas of this concept have been achieved, although a working quantum repeater has not yet been reported.

8.4   Summary

Most research in QKD concerns improvement of devices and raising the rates and distances currently achievable with today's technology. The vast majority of research up to now has employed quantum mechanical mechanisms, up to quantum computing. From a broad perspective, the security picture is more complex than simply producing quantum devices adequate for contemporary demands: quantum networks, as any network technology, will eventually be integrated into environments which are relatively weaker than the QKD building block within them. Much like a steel door in a wooden fence is useless if the adversary can simply ignore the unbreakable door and attempt the open window next to it, quantum networks can add strong security to the future internet, but alone they are insufficient in handling every type of security risk. MPT is one instance which illustrates how the strong building block "QKD" can be successfully combined with strong conventional, classic methods of accomplishing provable security at a level beyond basic key distribution. It remarkable to anticipate what other future applications can be built from QKD as strong primitives. Perhaps one of the most important benefits of QKD networks may not even be derived from its unbreakability in point-to-point connections but from its transparent and user-friendly establishment of key material for (symmetric) encryption. Today's conventional, i.e., intractable public-key cryptography, does not suffer from broken algorithms, but it fails in practice in almost all cases because of incorrect use. The effective management of public keys in daily business is a highly complex and nontrivial matter, requiring a deep understanding of mechanisms and good technical skills for creating, managing, distributing, revoking and using (public-key) certificates, public keys and private keys, and maintaining the proper refresh cycles. The latter also requires continuous monitoring of the requirements of public-key parameters (e.g., key lengths, admissible elliptic curves, etc.). QKD possesses the striking ability to substantially ease these matters by providing a silent and transparent source of shared secrets for encryption. Although this relates to symmetric cryptography (only), confidentiality, integrity and even digital-signature-like authenticity can be accomplished with the use of symmetric primitives only (see [51]). Most classic schemes from the symmetric paradigm already count as post-quantum secure. Remember that Shor's algorithm mainly endangers schemes based on factorization and discrete logarithms, and Grover's algorithm, as the only candidate which generically allows symmetric schemes to be attacked by brute force search, can be evaded through a humble doubling of key lengths to regain security at the same level which an attack without the use of a quantum computer would have. Simplifying key management by removing as much potential for human error in this regard is an additional, non-technical yet no less important benefit which QKD provides. This should not be ignored.

References

1. Ahmad, I., Shahabuddin, S., Kumar, T., Okwuibe, J., Gurtov, A., & Ylianttila, M. (2019). Security for 5G and beyond. *IEEE Communications Surveys and Tutorials, 21*(4), 3682–3722. ISSN 1553877X. https://doi.org/10.1109/COMST.2019.2916180.
2. Press, C. (2014). *Connecting Networks Companion Guide: Hierarchical Network Design.* Technical report.
3. Larsen, L. M. P., Checko, A., & Christiansen, H. L. (2019). A survey of the functional splits proposed for 5G mobile crosshaul networks. *IEEE Communications Surveys and Tutorials, 21*(1), 146–172. ISSN 1553877X. https://doi.org/10.1109/COMST.2018.2868805
4. Al-Obaidi, R., Checko, A., Holm, H., & Christiansen, H. (2015). Optimizing cloud-RAN deployments in real-life scenarios using microwave radio. *2015 European Conference on Networks and Communications, EuCNC 2015* (pp. 159–163). https://doi.org/10.1109/EuCNC.2015.7194060

5. Bjømstad, S., Chen, D., & Veisllari, R. (2018). Handling delay in 5g ethernet mobile fronthaul networks. In *2018 European Conference on Networks and Communications (EuCNC)* (pp. 1–9). IEEE.
6. Agiwal, M., Roy, A., & Saxena, N. (2016). Next generation 5G wireless networks: A comprehensive survey. *IEEE Communications Surveys and Tutorials, 18*(3), 1617–1655. ISSN 1553877X. https://doi.org/10.1109/COMST.2016.2532458
7. Dahlman, E., Parkvall, S., & Skold, J. (2020). *5G NR: The next generation wireless access technology*. Academic.
8. Zaidi, A., Athley, F., Medbo, J., Gustavsson, U., Durisi, G., & Chen, X. (2018). *5G physical layer: Principles, models and technology components*. Elsevier Science. ISBN 978-0-12-814579-1.
9. Jaber, M., Imran, M. A., Tafazolli, R., & Tukmanov, A. (2016). 5G backhaul challenges and emerging research directions: A survey. *IEEE Access, 4*, 1743–1766. ISSN 21693536. https://doi.org/10.1109/ACCESS.2016.2556011
10. Buchmann, J., & Ding, J. (Eds.). (2008). *Post-quantum cryptography*. Lecture Notes in Computer Science, vol. 5299. Springer.
11. Cho, J. Y., & Sergeev, A. (2021). Post-quantum MACsec in ethernet networks. *Journal of Cyber Security and Mobility, 10*(1), 161–176. ISSN 22454578. https://doi.org/10.13052/jcsm2245-1439.1016
12. Cho, J. Y., Sergeev, A., & Zou, J. (2020). Securing ethernet-based optical fronthaul for 5G network. *Journal of Cyber Security and Mobility, 9*(1), 91–110. ISSN 22454578. https://doi.org/10.13052/JCSM2245-1439.913
13. Cho, J. Y., Sergeev, A., & Zou, J. (2019). Securing ethernet-based optical fronthaul for 5g network. In *Proceedings of the 14th International Conference on Availability, Reliability and Security* (pp. 1–6).
14. Fröhlich, B., Dynes, J. F., Lucamarini, M., Sharpe, A. W., Yuan, Z., & Shields, A. J. (2013). A quantum access network. *Nature, 501*(7465), 69–72.
15. Zavitsanos, D., Ntanos, A., Giannoulis, G., & Avramopoulos, H. (2020). On the QKD integration in converged fiber/wireless topologies for secured, low-latency 5G/B5G fronthaul. *Applied Sciences, 10*(15), 5193. ISSN 2076-3417. https://doi.org/10.3390/app10155193
16. Milovančev, D., Vokić, N., Laudenbach, F., Pacher, C., Hübel, H., & Schrenk, B. (2021). High rate CV-QKD secured mobile WDM fronthaul for dense 5G radio networks. *Journal of Lightwave Technology, 39*(11), 3445–3457.
17. Eriksson, T. A., Hirano, T., Puttnam, B. J., Rademacher, G., Luís, R. S., Fujiwara, M., Namiki, R., Awaji, Y., Takeoka, M., Wada, N., & Sasaki, M. (2019). Wavelength division multiplexing of continuous variable quantum key distribution and 18.3 Tbit/s data channels. *Communications Physics, 2*(1), 1–5. ISSN 23993650. https://doi.org/10.1038/s42005-018-0105-5
18. Moghaddam, E. E., Beyranvand, H., & Salehi, J. A. (2021). Resource allocation in space division multiplexed elastic optical networks secured with quantum key distribution. *IEEE Journal on Selected Areas in Communications, 39*(9), 2688–2700. ISSN 0733-8716. https://doi.org/10.1109/JSAC.2021.3064641
19. Atakora, M., & Chenji, H. (2018). A multicast technique for fixed and mobile optical wireless backhaul in 5G networks. *IEEE Access, 6*, 27491–27506. ISSN 2169-3536. https://doi.org/10.1109/ACCESS.2018.2832980
20. Jaber, M., Lopez-Martinez, F. J., Imran, M. A., Sutton, A., Tukmanov, A., & Tafazolli, R. (2018). Wireless Backhaul: Performance modeling and impact on user association for 5G. *IEEE Transactions on Wireless Communications, 17*(5), 3095–3110. ISSN 15361276. https://doi.org/10.1109/TWC.2018.2806456
21. Cox, J. H., Chung, J., Donovan, S., Ivey, J., Clark, R. J., Riley, G., & Owen, H. L. (2017). Advancing software-defined networks: A survey. *IEEE Access, 5*, 25487–25526. ISSN 2169-3536. https://doi.org/10.1109/ACCESS.2017.2762291

22. Wang, R., Wang, Q., Kanellos, G. T., Nejabati, R., Simeonidou, D., Tessinari, R. S., Hugues-Salas, E., Bravalheri, A., Uniyal, N., Muqaddas, A. S., Guimaraes, R. S., Diallo, T., & Moazzeni, S. (2020). End-to-end quantum secured inter-domain 5G service orchestration over dynamically switched flex-grid optical networks enabled by a q-ROADM. *Journal of Lightwave Technology, 38*(1), 139–149. ISSN 0733-8724. https://doi.org/10.1109/JLT.2019.2949864

23. Wright, P., White, C., Parker, R. C., Pegon, J.-S., Menchetti, M., Pearse, J., Bahrami, A., Moroz, A., Wonfor, A., Penty, R. V., Spiller, T. P., & Lord, A. (2021). 5G network slicing with QKD and quantum-safe security. *Journal of Optical Communications and Networking, 13*(3), 33. ISSN 1943-0620. https://doi.org/10.1364/JOCN.413918

24. Aguado, A., Hugues-Salas, E., Haigh, P. A., Marhuenda, J., Price, A. B., Sibson, P., Kennard, J. E., Erven, C., Rarity, J. G., Thompson, M. G., Lord, A., Nejabati, R., & Simeonidou, D. (2017). Secure NFV orchestration over an SDN-controlled optical network with time-shared quantum key distribution resources. *Journal of Lightwave Technology, 35*(8), 1357–1362. ISSN 07338724. https://doi.org/10.1109/JLT.2016.2646921

25. Hugues-Salas, E., Ntavou, F., Gkounis, D., Kanellos, G. T., Nejabati, R., & Simeonidou, D. (2019). Monitoring and physical-layer attack mitigation in SDN-controlled quantum key distribution networks. *Journal of Optical Communications and Networking, 11*(2), A209–A218. ISSN 19430620. https://doi.org/10.1364/JOCN.11.00A209

26. Aguado, A., Lopez, V., Brito, J. P., Pastor, A., Lopez, D. R., & Martin, V. (2020). Enabling quantum key distribution networks via software-defined networking. In *2020 24th International Conference on Optical Network Design and Modeling, ONDM 2020.* https://doi.org/10.23919/ONDM48393.2020.9133024

27. Lo, H.-K., Curty, M., & Qi, B. (2012). Measurement-device-independent quantum key distribution. *Physical Review Letters, 108*(13), 130503. ISSN 0031-9007. https://doi.org/10.1103/PhysRevLett.108.130503

28. Tamaki, K., Lo, H. K., Fung, C. H. F., & Qi, B. (2012). Phase encoding schemes for measurement-device-independent quantum key distribution with basis-dependent flaw. *Physical Review A - Atomic, Molecular, and Optical Physics, 85*(4), 1–17. ISSN 10502947. https://doi.org/10.1103/PhysRevA.85.042307

29. Xu, F., Curty, M., Qi, B., & Lo, H. K. (2015). Measurement-device-independent quantum cryptography. *IEEE Journal of Selected Topics in Quantum Electronics, 21*(3). ISSN 21910359. https://doi.org/10.1109/JSTQE.2014.2381460

30. Calsamiglia, J., Barnett, S. M., & Lütkenhaus, N. (2001). Conditional beam-splitting attack on quantum key distribution. *Physical Review A, 65*(1), 012312. American Physical Society. https://doi.org/10.1103/PhysRevA.65.012312

31. Schartner, P., & Rass, S. (2010). Quantum key distribution and Denial-of-Service: Using strengthened classical cryptography as a fallback option. In *In Computer Symposium (ICS), 2010 International* (pp. 131–136). IEEE.

32. Rass, S., & Schartner, P. (2011). Information-leakage in hybrid randomized protocols. In J. Lopez & P. Samarati (Eds.), *Proceedings of the International Conference on Security and Cryptography (SECRYPT)* (pp. 134–143). SciTePress—Science and Technology Publications. ISBN 978-989-8425-71-3.

33. Fei, Y.-Y., Meng, X.-D., Gao, M., Wang, H., & Ma, Z. (2018). Quantum man-in-the-middle attack on the calibration process of quantum key distribution. *Scientific Reports, 8*(1), 4283. Nature Publishing Group. ISSN 2045-2322. https://doi.org/10.1038/s41598-018-22700-3

34. Acin, A., Brunner, N., Gisin, N., Massar, S., Pironio, S., & Scarani, V. (2007). Device-independent security of quantum cryptography against collective attacks. *Physical Review Letters, 98*(23), 230501. arXiv: quant-ph/0702152. ISSN 0031-9007, 1079-7114. https://doi.org/10.1103/PhysRevLett.98.230501

35. Zhou, L., Sheng, Y.-B., & Long, G.-L. (2020). Device-independent quantum secure direct communication against collective attacks. *Science Bulletin, 65*(1), 12–20. ISSN 2095-9273. https://doi.org/10.1016/j.scib.2019.10.025

36. Pironio, S., Acín, A., Brunner, N., Gisin, N., Massar, S., & Scarani, V. (2009). Device-independent quantum key distribution secure against collective attacks. *New Journal of Physics, 11*(4), 045021. ISSN 1367-2630. https://doi.org/10.1088/1367-2630/11/4/045021

37. Rass, S., & Schartner, P. (2020). Authentic quantum nonces. In C. Kollmitzer, S. Schauer, S. Rass, & B. Rainer (Eds.), *Quantum random number generation: Theory and practice.* Quantum Science and Technology (pp. 35–44). Cham: Springer International Publishing. ISBN 978-3-319-72596-3. https://doi.org/10.1007/978-3-319-72596-3_3

38. Gottesman, D., Lo, H., Lutkenhaus, N., & Preskill, J. (2004). Security of quantum key distribution with imperfect devices. In *International Symposium onInformation Theory, 2004. ISIT 2004. Proceedings.* (p. 136). https://doi.org/10.1109/ISIT.2004.1365172

39. Lo, H. (2017). Battling with quantum hackers. In *2017 Conference on Lasers and Electro-Optics (CLEO)* (pp. 1–1).

40. Vazirani, U., & Vidick, T. (2017). Robust device independent quantum key distribution. In *Proceedings of the 5th Conference on Innovations in Theoretical Computer Science*, ITCS '14 (pp. 35–36), New York, NY, USA: Association for Computing Machinery. ISBN 978-1-4503-2698-8. https://doi.org/10.1145/2554797.2554802

41. Cao, L., Luo, W., Wang, Y. X., Zou, J., Yan, R. D., Cai, H., Zhang, Y., Hu, X. L., Jiang, C., Fan, W. J., Zhou, X. Q., Dong, B., Luo, X. S., Lo, G. Q., Wang, Y. X., Xu, Z. W., Sun, S. H., Wang, X. B. ... & Liu, A. Q. (2020). Chip-based measurement-device-independent quantum key distribution using integrated silicon photonic systems. *Physical Review Applied, 14*(1) . ISSN 23317019. https://doi.org/10.1103/PhysRevApplied.14.011001

42. Cao, Y., Li, Y. H., Yang, K. X., Jiang, Y. F., Li, S. L., Hu, X. L., Abulizi, M., Li, C. L., Zhang, W., Sun, Q. C., Liu, W. Y., Jiang, X., Liao, S. K., Ren, J. G., Li, H., You, L., Wang, Z., Yin, J., Lu, C. Y., ... & Pan, J. W. (2020). Long-distance free-space measurement-device-independent quantum key distribution. *Physical Review Letters, 125*(26), 1–14. ISSN 10797114. https://doi.org/10.1103/PhysRevLett.125.260503

43. Wei, K., Li, W., Tan, H., Li, Y., Min, H., Zhang, W. J., Li, H., You, L., Wang, Z., Jiang, X., Chen, T. Y., Liao, S. K., Peng, C. Z., Xu, F., & Pan, J. W. (2020). High-speed measurement-device-independent quantum key distribution with integrated silicon photonics. *Physical Review X, 10*(3), 31030. ISSN 21603308. https://doi.org/10.1103/PhysRevX.10.031030

44. Wolf, R. (2021). *Quantum key distribution.* Lecture Notes in Physics (Vol. 988). Cham: Springer International Publishing. ISBN 978-3-030-73990-4. https://doi.org/10.1007/978-3-030-73991-1

45. Yin, H.-L., & Chen, Z.-B. (2019). Coherent-state-based twin-field quantum key distribution. *Scientific Reports, 9*(1), 14918. arXiv: 1901.05009. ISSN 2045-2322. https://doi.org/10.1038/s41598-019-50429-0

46. Chou, C.-W., Laurat, J., Deng, H., Choi, K. S., Riedmatten, H., Felinto, D., & Kimble, H. J. (2007). Functional quantum nodes for entanglement distribution over scalable quantum networks. *Science, 316*(5829), 1316–1320.

47. Dür, W., Briegel, H.-J., Cirac, J. I., & Zoller, P. (1999). Erratum: Quantum repeaters based on entanglement purification. *Physical Review A, 60*(1), 725.

48. Dür, W., Briegel, H.-J., Cirac, J. I., & Zoller, P. (1999). Quantum repeaters based on entanglement purification. *Physical Review A, 59*(1), 169–181. https://doi.org/10.1103/PhysRevA.59.169

49. Li, Z.-D., Zhang, R., Yin, X.-F., Liu, L.-Z., Hu, Y., Fang, Y.-Q., Fei, Y.-Y., Jiang, X., Zhang, J., Li, L., Liu, N.-L., Xu, F., Chen, Y.-A., & Pan, J.-W. (2019). Experimental quantum repeater without quantum memory. *Nature Photonics, 13*(9), 644–648. ISSN 1749-4893. https://doi.org/10.1038/s41566-019-0468-5

50. Yuan, Z.-S., Chen, Y.-A., Zhao, B., Chen, S., Schmiedmayer, J., & Pan, J.-W. (2008). Experimental demonstration of a BDCZ quantum repeater node. *Nature, 454*(7208), 1098–1101. ISSN 0028-0836, 1476-4687. https://doi.org/10.1038/nature07241

51. Rass, S. (2018). Perfectly secure communication, based on graph-topological addressing in unique-neighborhood networks. *Preprint arXiv:1810.05602.*

Printed in the United States
by Baker & Taylor Publisher Services